The Unlikely Journey

IT'S NOT JUST ABOUT THE HAT

"Life is good when you can dig in the dirt!"

Dr. Allan Armitage

4/21/23

First paperback printing, 2015

Library of Congress Cataloging-in-Publication Data
Armitage, Allan 1946 -
It's Not Just About The Hat / Dr. Allan Armitage.
p. cm.
ISBN 978-0-692-40779-0 (paperback : alk. paper)

Visit our website: http://www.allanarmitage.net/

Photo Credits:
Dennis McDaniel: Front Cover
Graetz Bros Ltd: Page 40
Meister Media Worldwide: Pages 73, 107
Judy Laushman: Pages 99, 130
Heather Dunagan: Page 142
All other photos provided by Allan Armitage.

PRAISE FOR
DR. ALLAN ARMITAGE AND
THE UNLIKELY JOURNEY OF A PLANTSMAN

"Love of life and family illuminates the paths Allan walks and the forks he takes. What a destination! Respected—dare I say revered—plantsman, innovative teacher, and treasured friend. The hat doesn't make this man—he makes the hat."

— Lis Friemoth
County Horticulturist, Blogger and Writer, The Garden Hoe, Wisconsin

"Dr. Allan Armitage's memoir, It's Not Just About The Hat, is like a fascinating, leisurely walk through a foreign, yet familiar, botanical garden; rich, colorful, and layered. To read the trajectory of this man who was first my teacher, and now my dear friend, is like stepping into another world—a world where barberry plants, hostas, wolves, talc powder, the open sea, and two-foot tall newborns illustrate and mold one of the most influential horticulturists of the 21st century. This candid Twainian account of his life brings insight to the man and the personality beneath the hat."

— Shanna T. Jones
Horticulturist and Marketing Specialist, Michigan

"As a student of Dr. Armitage all I knew of were the books, trial gardens, plant introductions, and the hundreds of students he loved to teach. Little did I know I was only scraping the surface! This was an eye opening read into experiences that developed the professor who changed my life. As per usual, the stories read with Armitage's style and flare. Dr. A's writing provides an informative yet captive read—from the lows of losing loved ones to the highs of becoming one of the most renowned professors in the country."

— Robby Jourdan
Project Manager, James Greenroofs, Colbert, Georgia

"Under the iconic Tilley hat is a true plantsman—a plant wizard. This delightful journey takes you along Dr. A's entertaining and accomplished career, careening through his garden path of life."

— Liz Klose
Director, MUN Botanical Garden, Newfoundland

AUTHOR'S NOTE TO READERS

I had no intention of writing about myself. I could not imagine anything more boring or presumptuous. However, after weaving many tales about my family and my background through plant talks and stories, people kept asking, "How did you get here from there?

When my good friends Rick Watson and Ed Kiley from the Perennial Farm persisted in asking that question, I finally decided to write a few things down. I am not a politician, a rock star, or a famous athlete, so I cannot enthrall you with drugs, sex, or money—just a few stories that somehow got me here from there.

— Dr. Allan Armitage

ABOUT THE AUTHOR

Dr. Allan Armitage is well known as a writer, speaker, and researcher throughout the world. Born and raised in Canada, he lived in Quebec and Ontario. He relocated to East Lansing, Michigan, and settled in Athens, Georgia. He has worked with landscape plants and greenhouse crops in both Northern and Southern climes. Allan travels from coast to coast, all parts in between, and throughout the world, sharing his passion for plants. He holds his B.Sc. from McGill University in Montreal, M.Sc. from the University of Guelph, and Ph.D. from Michigan State University.

Visit Dr. Armitage's website, www.allanarmitage.net to learn more about Allan's recent antics, upcoming tours, adventures, books, and apps to promote the exciting world of horticulture.

It's Not Just About The Hat

DEDICATION

To My Incredible Wife Susan.
Without Her, I Would Be Lost

CONTENTS

ACKNOWLEDGEMENTS

FOREWORD

PROLOGUE

AFTERWORD

ACKNOWLEDGEMENTS

My journey has been great fun to share, but would have not occurred without the help of many people, especially the following: Susan Armitage, Dawn Hummel, Shanna Jones, Jennifer Koester, Ed Kiley, Tracy McPherson, Rick Watson, and Maria Zampini.

"The difference between stupidity and genius is that genius has its limits."
— Albert Einstein

"It is better to wear out than rust out."
— Richard Cumberland

"When you come to a fork in the path, take it."
— Yogi Berra

FOREWORD

Allan has outdone himself with his latest accomplishment. This remarkable book is a must read and unlike any other personal account of a journey through life. It is not just about a journey—it is about a life filled with adventure and discovery. As Allan shares his adventures with his readers, we feel we are travelling with him. He takes us from playing and coaching sports to mixing up different soup recipes for Del Monte. He brings us along on a North Atlantic crossing on a merchant ship; he seeks his fortune in Europe and eventually finds his professional passion: horticulture.

When I picked the book up I couldn't put it down. It is skillfully written with a rare mix of loving attention, humor, practical wisdom and is chock full of life's experiences. Allan intermingles stories about his childhood, family and professional accomplishments. His heartfelt stories are funny, insightful, and inspiring. Allan's adventure takes us from his humble beginnings in Mount Royal, Canada, to Lansing, Michigan, and then on to Athens, Georgia, where he gained great notoriety as the foremost horticulturist and lecturer of his time.

A warm and glowing book that I will share with my personal and professional friends.

— Ed Kiley, Director of Sales and Marketing
The Perennial Farm

PROLOGUE

There I was, sitting across from Nicole from the QVC television network. I had just spent the entire day being trained for a possible guest appearance to sell plant food on the show. I had completed an online test a few weeks earlier, and then flew to the studios to be certified. Around a table were twelve aspiring pitchmen and women hoping to sell their wares. The staff evaluated our spiels and provided advice on television salesmanship. I had just recorded two "real" practice sessions under the lights in front of cameras. Each one was critiqued and appraised. My coach looked up at the end and said, "Congratulations, you have passed the certification portion," and shook my hand. "The last thing you need to do is visit Nicole." It turned out that Nicole was an image consultant at QVC. It also turned out that not everyone was invited to see her, only a chosen few.

No one ever accused me of being a model for Calvin Klein, but I believed I had dressed reasonably well. I was sporting a jacket, my favorite corduroy pants, and an off-white golf shirt—it even had a collar. Nicole eyed me up and down and got down to business. "That corduroy jacket, get rid of it. You are selling gardening, springtime, and color. It has to go". "And those pants...you can wear jeans, but they need to be better fitted and dark blue, not stone-washed or colored." She had lost me at stone-washed but went on to disparage my entire wardrobe, even down to my running shoes. "Those shoes—you need a new pair." I retorted, "But I am behind a table, no one even sees my shoes." She finished her sacking by leaning back, looking at me for a full minute, and declaring, "Chambray. You would look good in chambray." Needless to say, I had no idea what chambray was, nor did I have any idea why it would look good on me.

It was then that it hit me again. What is a boy from Montreal doing in a television studio in Philadelphia? *How did I get here from there?*

One

"Remember when we were kids and we couldn't wait to grow up.... What were we thinking???"

. . . .

- Anonymous

I must start at the beginning, when I met my first hedge....

July, 1951

"Geez, that really hurts. What did I do anyway?" "Nothing. You were just bugging me." So transpired my introduction to plants; my older brother Howard, whom everyone called Hasie, had just tossed his five-year-old brother into our barberry hedge—and I learned that barberries have very sharp thorns.

I grew up in the town of Mount Royal, affectionately known as TMR, an emerging suburb of Montreal. The Fifties were a more innocent time; we played in the fields behind our semi-detached house, and we walked a mile or so to school or to the rec center to play hockey and baseball. We lived a rather "normal" life, or so I thought. My dad was a big man with a big voice who seemed to change jobs a good deal. During his time at the Montreal Stock Exchange, he would yell orders all day, then come home and do the same. And when he came to our ball games, you could hear him cheering from home plate to center field. He was a good man in many respects, but a little scary. He loved to take Sunday rides in the car and ask if anyone wanted to join him. Hasie was always ready to go; I seldom went. They travelled many miles and came back in fine spirits.

Allan and Hasie, 1987

To this very day, I do not know why I did not join them. It is one of the few regrets of my boyhood. I think I may have been a mama's boy.

Like all the mothers in the neighborhood, Mom stayed at home taking care of her unruly sons and rather unruly house. She was a great mother. Very bright, and good at lots of things, but keeping a clean house was not one of them. We brought in the empty trash cans and helped with the dishes, but we were of little help in the tidiness department. My brother and I were totally oblivious to the off-color curtains, the somewhat dirty windows, the occasional scraps of material that littered the floor, or the layers of talcum powder in the bathroom. Mom loved her talcum powder. She simply could not keep it away from every surface in the bathroom. Such minor detritus was normal, but it was tough on Dad. He liked his shirts starched, his handkerchiefs folded and his shoes put away.

. . . .

Little did we know we were living with an urban legend! The house was, as poets put it, a little unkempt. To be sure, it was messy, but with two rowdy boys, it seemed pretty ordinary. It turns out that the phrase "ignorance is bliss" could well have been invented by Hasie and me. It was not until we were teenagers and brought friends home that we were even aware that the place might not quite be up to snuff. We seldom brought girlfriends home, such relationships never being of sufficient depth that we would want them to meet our parents.

The living room was bright and airy. That the white curtains were yellow never registered with us, until Hasie finally brought his girlfriend home and she commented on them. The furniture was also kind of grubby. Cigarette burns in the Persian carpet should have been a hint that June Cleaver did not live there. We did not know anything of demographics, but it turned out that our parents were the last of the hard-drinking, hard-

smoking, post-depression, post-war breed. Du Maurier cigarette stubs and liquor bottles were as common around our house as the tomato and rice casserole we ate every other night. To this day, I remember those blackened divots in the carpet. Now, when I see a Persian rug, I think it must be missing something. The furniture was similarly scarred and acted as excellent receptacles of cigarette ash. This just seemed normal. It is a miracle we did not all go up in smoke.

Going to the kitchen was an adventure. We were never sure what might greet us on the countertop or in the cupboards. They may be a total mess, or they may be just a little untidy. However, the one thing that never changed was the gin bottle in the refrigerator. There it was, full of cold, refreshing water. Always! To say it was grungy was to be kind; it had been refilled for countless weeks and months, perhaps years, and harbored a brownish cast. When we were thirsty, we removed the cap and helped ourselves to the bottle, swigging sufficient volumes to quench our thirst—then put it back. When it got low, we simply topped it up. I doubt my mother was part of the swigging, but her three men never thought twice about it. Didn't everyone keep a gin bottle of cold water in the fridge? That ours looked like a brackish swamp in a cruddy old bottle didn't seem unusual. It is one of the many visual moments that we recall when my brother, myself, and our wives get together. Howard's wife, Phyllis, looks at him with a "How could you drink from that thing?" manner. He just glances at me and replies with a "Who knew?" shrug of his shoulders. If that gin bottle were around today, it would be in the Smithsonian. Actually, if it were around today, it would bear a skull and crossbones label.

My mother had many friends. She could talk on the phone for hours. Little did we realize it, but the phone was another potential Smithsonian treasure. It was a typical green rotary phone of its day, but its color seemed to have faded. The fact is, it was just plain scuzzy. Its soft green hue was mottled with soft brown grunge—and like veins on a leaf, was littered with tracks from idle fingernails. It sat in a small alcove buried in the wall in the hallway, its cord to the earpiece a hellish maze of twisted, knotted wire. Although the cord was about ten feet long, we had to scrunch close because we couldn't pull at the earpiece without dragging the phone out of its perch.

As most houses did back then, ours had a single bathroom. The bathroom was always a bit of a problem, but it was a scene out of the Dust Bowl when my mother decided to apply talcum powder to her face and body. There must have been little gremlins gleefully throwing the stuff all over her body, as there was white powder everywhere. The sink, the floor, the toilet seat, everywhere, every day. Dad would get quite upset so she cleaned the place as well as she could, which is to say only a thin layer remained over everything. When she moved to her own apartment many years later, she must have taken the gremlins and all their relatives with her. My kids still talk about the footprints of talc from bathroom to bedroom, and the chalky hue of the bathroom when they visited her there.

. . . .

From these accounts, the reader may think our family a bit strange. You may even feel a little sorry for the boys. Nothing could be further from the truth. To Hasie and me, all of these circumstances were as normal as Dad spooning bloody juice from the Sunday roast into our waiting mouths. Didn't everyone do that?

The Characters

Our semi-detached house sat on a sixty-by-one-hundred-feet lot on Melbourne Avenue in TMR. It was a rather typical yard, far more utilitarian than ornamental. The landscape consisted of the barberry hedge in the front, a few junipers around the front door, grass in the front yard, and not much else. There was no hedge or other barrier between our house and that of our attached neighbors, the Smiths. This was a very good thing, in that our yard doubled in size and quickly became the football field in the fall. We tried to be good neighbors, but there was no doubt the boys and their games were doing little to improve the turf.

The Smith boys, Ronnie and Dougie, were twins. They were the only practicing Catholics in the neighborhood. They went to a Catholic school and spent a lot of time studying catechism. Unfortunately for us, they were not able to spend a lot of time with the neighborhood kids. But their lawn did. Mr. and Mrs. Smith let us know every now and then that they would appreciate us letting their grass grow—at least a little. They were very patient and never put up a fence or hedge.

However, patient they were not when we got bored inside the house and threw tennis balls against our dining room wall—the common wall to both houses. Thump, thump, thump—it must have driven them crazy! After a few too many thumps, they would call or knock on the door after Dad got home. After a long day at work, our father did not need complaints about his sons and said sons knew nothing good would come of that. We did not let that happen very often!

The side of the house sported a large honeysuckle hedge, the kind with dark green leaves and bright red berries in the summer. It was at least six feet tall, monstrous to us kids. It served many functions, not the least of which was an armory. The berries made wonderful ammunition. Every now and then, the boys in the neighborhood would have friendly pitched battles. Our friends' mothers, not to mention our own, were not at all pleased as they attempted to scrub those noxious scarlet stains off their boys' clothes. The other advantage of the hedge was that it served as a goal to practice dropkicking a football. Few boys today know the term "dropkick", but that was the preferred method of youthful point-after touchdowns in our days. From our driveway, we practiced kicking the ball over the hedge. We got pretty good at it. Unfortunately Mr. Whitby, who lived on the other side of the hedge, was not impressed.

Allan around eight years old, putting on his life vest

Canada is a comparatively young country, made up mostly of English speakers, but Quebec was predominantly French, even in the 1950s. However, many parts of Montreal were not only English speaking, but were very "English." Canada was part of the Commonwealth. We were tied to England politically and philosophically, and many of our neighbors were as English as tea and crumpets. Just down the street lived my good friend Norman Bengough—otherwise known as Benny. Every summer evening, we would hear "Nooo-maan" at least three times as his very English mother called him for dinner. I can hear her to this day.

Mr. Whitby was also as English as the BBC. Perhaps he enjoyed cricket, but he was not amused with cacophonous kids playing ball or with footballs landing on his lawn or stairway. His son Rigby was our age, and we got along fine. To no surprise, he did not join us in kicking balls over the hedge. Mr. Whitby had a few conversations with Dad as well. We were not making many friends with our immediate neighbors.

Learning About Plants

We did not have much in the way of a garden. We had "yards" in the front and back. The barberry hedge in the front, the honeysuckle hedge along the side, a shade tree near the street, a few green things by the foundation, and our functional lawn in the front. Along the side of our garage was a small path leading to the back. The backyard was grass surrounded by a cedar hedge. English though our ancestors were, I don't recall many "gardens," although, at that age, I wouldn't have known one if I tripped over it. The gardening gene seemed to have skipped my parents, but then again, we did have the hollyhocks.

Hidden away on the side of the garage, they were Mom's favorite flowers. Each spring they emerged from the ground and like Jack's beanstalk they grew taller and taller. Attention to garden detail was not the norm in the Armitage household, but in the case of the hollyhocks, I was tasked to run strings from nails embedded in the brick wall and tie up the stalks as they developed. The plants were green, the brick was red, and the string was white. Who needed flowers? Each year, as the season progressed, I wound the strings around the stalks and each year, I watched bugs eat holes in the plants and rusty brown spots disfigure the leaves. Jack's beanstalk they were not. Although I did not know it, I was learning about Japanese beetles and plant rust. To this day, whenever I see hollyhocks—regardless of their beauty—I think back to those unfortunate bug-laden, fungus-ridden flora, and I am nine years old once more.

One of the bedrocks in our life was our grandmother on Mom's side, known simply as Gorba. I remember many things about her house—it was always clean for one thing, and she cooked the best meals. Hasie and I often clambered to a small room at the top of the stairs where she kept books about different countries. I would love to look through them and ask about places like France and Italy. I was too young to remember any specifics, but one of the things I do remember her saying was, "Travel is a wonderful thing; no one can ever take those memories from you." It must have been Gorba who gave me the travel bug.

I like to think of myself as a positive person, optimistic about the future and a person who believes in people. This optimism probably came from Gorba. When she passed away, my brother noted the same trait in her eulogy.

· ·

When it came to family members and grandchildren, Gorba was horribly naïve. None of us could do any wrong. All I knew was that when I left her place, I had a renewed sense of value. She had the most unbelievable way of making me feel important. She got me through some tough times and gave me value—to know I was special.

· ·

I also remember her garden. To get to her back yard, we took a shady narrow walkway beside the garage. That was the first time I noticed a variegated plant, thriving in that environmental wasteland. I remember walking down the path where variegated hostas had long ago been planted on either side. It was a dim, dark pathway but the light airy foliage lit the way. Nothing else would grow there. That path was a living testament to the toughness of these plants. Gorba called them funkias—the old name for hosta—and they made my grandmother seem like a horticultural genius.

Much later, we would sit under her large catalpa tree, sip iced tea, and catch up with each other's lives. As refreshing as the tea was, I constantly looked at the long green icicle-like fruits of the catalpa, wondering upon whose head they would fall. When my wife Susan first met Gorba, she was as enamored with her as I was. She was a rock throughout my life and lived in her house until well into her eighties. Gorba passed away in 1988, at the age of ninety-five.

Gorba at Allan's wedding

Two

"I get to go to overseas places, like Canada."

. . . .

- Britney Spears

Hanging Out in Montreal

We were not poor. Mom could always find a dime or quarter if we needed one. Unfortunately, Dad seemed to lose his job every now and then. Even though we had paper routes and other odd jobs, there was little enough money for extras. Spending time with my brother was about the best thing in my entire life. Even though he would get upset with me every now and then and pummel me into submission, he was still my hero. I recall when Hasie would walk with me to town, about a mile or so from the house, to just hang out. We sometimes walked into our little drug store/lunch counter and propped ourselves up on the stools. We would sit up on the rotating stools, nod to the locals drinking coffee and order lettuce sandwiches. Two slices of Wonder bread and a leaf of lettuce—five cents apiece. That was all we could afford. I am sure our waitress Janice took pity on us—a lettuce sandwich was not normal fare. But for me, sitting beside my big brother and tasting that sticky white bread on the roof of my mouth was Nirvana. To this day, I will say the heck with whole wheat and have a slice of that unhealthy but memorable white stuff.

As we got older, we did the things of boys. We played in vacant fields, we got into trouble by staying out, sleeping in, and avoiding our homework. But mostly, we played sports. Hasie was one of the best athletes around—he could do anything. He was the best baseball player in the rec leagues, could catch a football anywhere near him, and was a terrific hockey player. I was not half bad either—when we teamed up, we could beat anybody in almost all two-on-two sports.

In the spring, we practiced balls and strikes. Hasie would pitch, and I would be catcher and umpire until he struck out the side. Then we would switch. He was three years older, bigger, and stronger, and I learned not to call too many fastballs. None of us had any idea if we were particularly talented, as in Canada there were no college scholarships or fervent emphasis on sports as a future. We played in rec leagues, and later played in high school, but no matter how good anyone was, it was just a game. We competed hard, and we had fun, but no one ever dreamed of college scouts or college scholarships—they simply weren't in the language.

If you were a boy growing up in Canada—certainly in Montreal—you played hockey. Every pastime took a back seat to that game. We grew up with the likes of Bernie "Boom Boom" Geoffrion, Maurice "The Rocket" Richard, and Jacques Plante. Winter was long and snowy, and once descended, it stayed. We often played pickup hockey on the street, otherwise known as "shinny." The streets were always a little snowy underfoot. Snow drifts as tall as young boys were the norm. If a few boys got together, we dug a goal out of the snowdrifts on the side of the road and fired rock-hard tennis balls at the goalie, then switch off.

The best times, however, were when a bunch of us showed up on a Saturday morning, and a serious game of shinny evolved. These games often took place on our street, with pieces of ice defining the goalposts. A frozen tennis ball was the puck. Six, eight, ten boys would show up with their street sticks (used sticks not good enough to use on in a real game on ice), and we played for hours. This was before helmets, goalie masks, and mouth guards.

When I was around nine, I was the goalie during an intense game of shinny. Suddenly, a melee took place around my net. I went down to block an incoming shot. When I rose, everyone was staring at me with mouths open. Hasie immediately took me to the house, whereupon Mom opened the door, uttered a cry, and almost fainted. A four-inch wooden splinter had penetrated my nose and lodged there. I must have looked like an African warrior with a bone through his nose. At that point, it really did not hurt. There was no blood, but everyone sure took notice. If this had occurred today, iPhones would have snapped a dozen photos. When we walked into the hospital ER, even the nurses stared—nine-year-old boys with sticks through their noses were not common occurrences. Amazingly, it had not hit the cartilage or broken my nose, but as time went on, the pain started to take hold. And when the doctor broke off one end of the stick and pulled on the other to remove it— that smarted! Once removed, I received a couple of stitches in each side of my nose and was sent home. Even now, when I tell this story to skeptics, I need only to show the scars, and they become believers. Our nine-year-old grandkids think they are really neat.

The backyard was our personal rec center and evolved as a part of my burgeoning career in athletics. In the winter, we sometimes froze a small rink in the back yard and played there. As we got older, we would walk to the town rinks and play hockey there. During the summer, it seemed silly to waste a perfectly good yard with grass. My parents had purchased some steel drinking-glass holders, with a coiled stem that was plunged in the ground. The top was in a cylinder shape for a beer glass to be placed beside chairs for drinks outside. I don't ever recall my parents battling mosquitoes and black flies to sit around on our weedy lawn, but the holders did not go to waste. I stuck a couple in the ground against the back wall of the garage, mounted some pillows on the cylinders, and threw fast balls from the pitcher's area sixty feet away. It did nothing for Mom's pillows or for the baseballs thudding off the brick wall. But I sure learned to throw a strike.

One year we built a wooden basketball goal in the back yard. It was crude, poorly built and, with only fraying rope cables supporting it, was in constant danger of toppling over. I spent hours shooting hoops with friends on the rickety backboard. As you can imagine, the grass in the backyard did not do well. I often wonder if my parents were "gardeners," would they have allowed such a monstrosity to be built? Would green grass have superseded bouncing balls?

Aside from tending to the hollyhocks, our yard work consisted of occasionally trimming the hedges and cutting the grass. The more the backyard became an exercise area, the less grass there was to mow. However, we did have a few petunias in the back, as far away from the basketball court as possible. Mom would plant them against the cedar hedge, and I weeded them every now and then. To me, they were a total waste of space.

Not Exactly My Favorite Times

Other than seasonal allergies, bumps, bruises, and occasional disagreements with friends, it could be said that my childhood was rather boring. Then along came high school. There is no such thing as middle school in Quebec, so off we went at age thirteen to spend our formative teenage years in the Big House. Town of Mount Royal High School was a typical high school in English Quebec in the early 1960s, where the biggest vices were smoking, drinking, and usual teen hijinks. The Sixties were the decade of radical changes and protests in America, but not so much in Quebec. The driving age in Quebec was seventeen, so only some seniors in high school were old enough to drive. There we were, young children really, spending four years preparing for whatever was to come next. I was not particularly smart, but I did not have to work too hard either. I was an average shy skinny kid who, along with other average shy kids, sat in the middle rows—neither the class clowns nor the class achievers—just there. I do not remember many of my teachers, which is kind of a shame, but there were some good ones among them, such as Mr. Scammel. Arthur Scammel was from Newfoundland and was an excellent teacher. He was well known in Maritime Canada and at TMR High for writing the fishing ballad "The Squid Jiggin' Ground." It was recorded by Hank Snow and was the representative song for Newfoundland when the province joined Canada in 1949. For teenage kids, Mr. Scammel did not compete with the Beatles but was a bit of Canadiana we were fortunate to bump up against.

Have you noticed that the human psyche remembers bad events more routinely than good ones? I believe this holds true for bad teachers as well. I certainly recall some unpleasant ones, like the French teacher who taught by intimidation. He became apoplectic on hearing poor verb conjugation—and heaven help you if called upon on a day when your homework was not done. We all learned some French to be sure—I must admit I can still recall some of his lessons. Upon looking back, he was probably a man lacking in self-confidence. Perhaps he hoped to gain some at the expense of terrified fourteen year olds.

My high school years were not among my favorite times. Fortunately, I was a reasonably good athlete and stayed on the ball field, hockey rink, and basketball court, and out of trouble. I was a fairly talented basketball player, starting on the high school team, and we did quite well. I sometimes wonder how much better we would have been if not for the basketball coach. He was a caricature parody of a poor physical education teacher. He had to do something "academic" so was assigned to teach "health class" to teenage boys. His teaching methods consisted of calling on a kid, handing him a package of hand-me-down notes, and commanding him to write them on the blackboard. At this point, he would put his feet up, and read the day's paper. When the boy got tired—or ran out of chalk—he told another kid to continue. Our time was spent copying the notes, usually illegible, until the bell rang and we left.

If I only had to put up with that incredible incompetence once a week, that would have been fine, but he was also as poor a coach as I can recall. He ranted and raved at his players, teaching nothing while sapping talent and enthusiasm. If I should miss a shot or mess up a pass, he would yell out such endearing terms as, "Armitage, you pinhead!" He was eventually replaced with Mr. Wade, a far better coach, and with him, we won our fair share of games. But that was also because of Mr. Neal.

TMR Basketball Team, 1962
Back row: Coach Bill Wade, Philip Johnson, Pete Waye, John Howard,
Danny Watkins, Peter Conradi
Front row: Ian Gross, Allan Armitage, Allan Lanthier, Alex Patterson, Axel Conradi,
Brad Jones, John Harris

Bob Neal volunteered at the United Church, whose best attribute was its basketball court in the basement. He would be there while "his" teenage boys played ball every Friday night. He was everything our high school coach was not—encouraging, accommodating, and approachable. I recently attended my fiftieth high school reunion. I found Axel Conradi, one of my former teammates, for the first time in that many years. We both agreed that if not for Mr. Neal, we would have been typical teens in the bars on Friday nights. There is no doubt that my experiences with sports and with those teachers were huge influences in the next phases of my life.

Immersed In French

When I was in high school, my dad thought it would be a good idea to send his boys off to a farm in the Eastern Townships of Quebec during the summer. French was the first language of the province. Although we started studying the language in grade three, Dad felt that his boys would learn it far better on site, while building muscle and character. Hasie went first and like anything else, he excelled. He loved the farming routine, he enjoyed the Quebecois lifestyle, and, most importantly, he picked up French seemingly without effort. He did it for two summers, and perhaps, as a result, enrolled at MacDonald College—the Agricultural College of McGill University.

I hated it. Not only did it mess up my plans to be a professional baseball player, but I was not at all enamored with the farming lifestyle as my brother had been.

One of the things my father did not consider was my severe hay fever. As a kid, I often awoke unable to open my eyes because they were glued shut with gunk from allergies. My eyes were swollen and red. My throat itched, and my nose was a mess for at least six weeks in the summer. We threw hay bales onto a wagon, stacked the bales in a dusty and moldy barn, worked in the corn and tobacco fields, and hung tobacco in the barn. Every single one of these activities was like an opening bell for a boxing match that had already gone ten rounds. I was miserable.

I missed my routine and my girlfriend. To add to all of that, I was lousy in French. There is no doubt Hasie got all the language genes—he speaks French, Spanish, Portuguese, and some Swahili— and I, none. I did learn a great deal about cows, chickens, hay, tobacco, and corn. When not sneezing, I enjoyed being outdoors and getting back to the land. However, my inability to become conversant in French kind of defeated the entire purpose, and it made for long, long summers.

Hit the Brakes

Every now and then, Hasie would come home from college to check up on his little brother. This was a highlight for me, because I knew he would always take care of me—well, perhaps not always.

· ·

When I heard Hasie was coming home from college, I could hardly wait. I was so excited when he offered to teach me to drive that I left school early. When he arrived, away we went in the beige Ford Falcon. I was doing pretty well until I turned in the driveway to park the car. At that point all hell broke loose as the car continued to move towards the garage. Hasie kept saying, "Hit the brakes, hit the brakes!" but my brain went blank. I kept hitting the clutch instead. It was all over in a matter of seconds, at which time the garage door buckled under the impact. The car came to a stop. My loving, caring brother looked at me across the seat and said, "Oh boy, Dad is going to kill you." He extricated the car from the garage, reversed down the drive, and headed back to school. I fretted all day. There was no way one could hide a smashed garage door, and running away didn't seem like much of a plan. I was one frightened fourteen year old that evening. After a little shouting on Dad's part and sufficient remorse on mine, he suggested I should use my paper route money to help pay for the damages. "Yes, sir!" was all I said.

· ·

During my teen years, my parents' relationship continued to deteriorate. Perhaps they felt that with the eldest gone, one reason to be civil had disappeared. Marital discord often sent me to my friends' houses. Or I would lie on my bed, turn on the radio, and listen to music rather than the chaos below. Compared to the term today, I certainly was not a troubled teen, but the stress took a toll on all of us. Not long after I finished high school, they separated, and soon thereafter divorced.

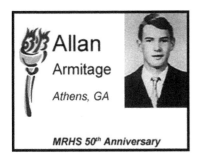

Allan
Armitage
Athens, GA

MRHS 50ᵗʰ Anniversary

Horse, 50th high school reunion, 2013

All in all, my high school years were rather routine. I played basketball and hockey, said no to cigarettes, yes to an occasional beer, and met girls. I hung out with friends like Itchy, Benny, Rats, Boo-Boo, Launch, and others. Everybody in those days had a nickname; mine was Horse. At our fiftieth high school reunion, I was welcomed as Horse. I greeted Moose and Launch once again.

Three

*"You can lead a boy to college,
but you cannot make him think."*

. . . .

- Elbert Hubbard

Look to Your Left, Look to Your Right

As high school came to a close, I knew I would attend college somewhere. I had absolutely no idea of my future—I only knew I needed to get out of the house and out of town. Some of my classmates were applying out of province, others to various universities in Montreal. I was seventeen years old when I graduated from high school. I had no clue as to where or what I would do with my life. But really, for this shy skinny kid, there was little doubt: I would follow my brother.

It was a poor excuse to choose an education, but it was a comfortable one. Applying to an agriculture school when I hated farming did not make a lot of sense. But it was a getaway, and it was convenient. Today, kids visit campuses all over the country. Parents accompany them with counsel and advice but for me, it was apply and go. I was the only one from my high school with any desire to go to an agricultural school, and that was fine with me. I was looking more for an escape than an education. Somehow, I was accepted. In September of 1963, I found myself twenty miles from home starting a new life.

McGill University was—and still is—one of Canada's most prestigious centers for higher learning. It is often referred to as the Harvard of the North, but Harvard should be so well respected. MacDonald College was a satellite school of McGill, located on the far west end of the island of Montreal, in the beautiful town of Sainte Anne de Bellevue. It was an English language academic institution located in a French blue-collar town. The campus was an island unto itself, with little interaction between the town and the students. It was also a small college, housing Agriculture, Physical Education, and Dietetics, then known as Home Economics. Fewer than 1,800 students were on campus, but its lack of size did not diminish the academic difficulty and intensity of interpersonal relationships. In short, it was a place where we grew up.

I was determined not to be that shy skinny kid anymore. I went out of my way to join in, party with, and hang out with as many people as possible. The course load was heavy and difficult. I studied comparative literature and math and tackled organic chemistry, biochemistry, physics, biophysics, higher math, and many sciences in the first two years.

The school year consisted of two terms. Grading was based on the usual percentage system of one to one hundred. However, if you failed more than one class in any term, you were gone. We were simply expected to attend classes, take notes, review notes every now and then, and take exams when necessary. We did not write exams during the term. The Christmas exams were considered midterms. Canadian universities at that time were loosely based on the English grading system. The final exams counted for one hundred percent of the grade. There were no little weekly quizzes, or "gimme" grades so prevalent in college courses today; it was pass or fail. All or nothing—once at Christmas and then again in the spring.

For immature kids, this was the worse system imaginable. Other than the midterms that we all knew counted for nothing, there were no speed bumps, yields, or stop signs to influence our behavior for four months. I am sure there were advisors back then, but I certainly was not aware of them. When the exams descended twice a year, it was like the Charge of the Light Brigade—lots of casualties and no decisive victories. I remember looking at the list of grades after Christmas exams the first year. I was really worried about an English course. I breathed a huge sigh of relief when I saw Armitage—fifty-five percent. I had passed! A grade of fifty percent was the Holy Grail for students like me. It meant we could fight another day. A grade of seventy percent was outstanding; a grade above eighty put you in the genius category. Given today's standards, one could look at students happy with a fifty-five percent and deduce that the grading system must have been easy. In fact, it was a very unforgiving system meant to weed out those less mature.

When we first arrived on campus, the dean said to the incoming freshman class, "Look to your right, look to your left—one of you will not be here next year." He was serious—and correct. We lost one third of our class following Christmas exams, and probably half the class was gone after the first year. By the end of the second year, of the original one hundred freshmen, only twenty-five remained.

Those who succeeded the first two years were not necessarily smart. We simply understood that we had to prioritize. I was no brighter than the next person. I skipped my fair share of classes to sleep in or play hockey during free ice times. I partied way more than I should have and was lucky to have made it through those first two years, academically and personally.

The local watering hole in town was known as Joe's Tavern. Even in the dead of winter, we would walk to town and stumble home. That we did not freeze to death was perhaps in part because of the ethanol stored in our bodies. In our second year, Joe's burned down. Our drinking had to be carried out across the bridge in the next town, about ten miles away. We would jump into cars in the middle of snow, ice, and sleet to cross the bridge. We partied on and crossed the river once again. This was before seatbelts, designated drivers, sobriety stops, and common sense. That we didn't crash through the bridge, run into a tree, or hurt anyone else was a miracle. I shudder now whenever I read of a DUI or some alcoholic tragedy. I have no doubt that someone was looking after us; we were too young and stupid to look after ourselves.

Saved by the Ball

Pat and Alma Baker

The toughness I learned on the court in high school and the encouragement from Bob Neal must have been useful. Pat Baker, the basketball coach at MacDonald, recruited me to play there. I use the term "recruitment" very loosely, as there were no athletic scholarships, home visits, or national rankings in those days.

However, he had seen me play and was more than pleased that I was to attend Mac. I played ball there all four years with great success. We practiced hard, our teams did well, and I personally received many accolades. I still visit Pat when I return to Montreal. He was without doubt one of the most positive people in my life. Lord knows what bridge I may have driven off if not for his teaching skills and encouragement. Not a single player I played with had anything bad to say about Pat. Every one of them still counts him as a mentor today. The world certainly needs more Pat Bakers. I was indeed fortunate.

We had some highly skilled players, competed against some excellent teams, and won more games than we lost. However, to put small college Canadian ball in the 1960s in perspective, we would have been hard pressed to beat a decent Division III team in the States. We were the smallest school in our division. We played in a gym built over a dining hall that seated about two hundred people, on a good night. Our strength conditioning consisted of walking around campus from class to class. During my entire four-year playing career, I believe Doug Boyd, at about six-feet-four inches, was our tallest player. He was an excellent ball player, but dunking was not a highlight in his repertoire. In fact, I recall very few of us routinely dunking the basketball. I still see Doug at shows every now and then and we reminisce fondly about those years.

I also played varsity football for two years. We worked equally hard and had some truly talented players. However, we would not have won any championships in the United States, to be sure.

As I observe the machinery that is currently college sports in the United States, I would not have changed a thing. Athletics provided healthy competition, a shared camaraderie, and reasons to work hard and to stay fit. Being part of a team built self-confidence that could not be duplicated in any other endeavor. We went to school to grow up. Athletics were simply one more aspect of our education, even though I had no idea what I was learning. It is said that college is a suspension of time from the real world; I couldn't agree more.

I went to college in the Sixties, and the courses were doubtless challenging and cutting edge. However, I am somewhat embarrassed to say that I remember very few of them. This was not because it was the Sixties and we were high on the newest drugs of the day. In fact, other than music, coming of age at that time in Quebec was no different from any other decade. Sexual morals were loosening up, but the girls in the dorm still had a curfew of eleven p.m. Beer was still the beverage of choice, and most kids who flunked out did so because they were lazy or played bridge in the cafeteria for too many

Mac basketball, 1967

hours. I don't know any students these days that even play bridge, let alone enough to affect their futures. And people think Canadians are boring...

We had classes every day, including Saturdays. The most trouble any of my friends got into was skipping too many classes because they would rather play hockey, figure out ways to get in the girls' dorm after eleven, or not hit the books. We partied hard like any young people those days. There were just fewer opportunities to get into trouble than there are today.

Canadian universities had no formal Greek system, so frat parties and tailgating were never part of our vocabulary. This is not to say we did not party with the best of them and carry on like thoughtless, irresponsible, poorly behaved young people every now and then. It is just that the worst of the Sixties simply passed us by. However, we knew times were changing. Television news brought us Timothy Leary and a substance called LSD. I saw young men opting out of a war they didn't believe in and coming into Canada. And not one hundred miles south, I saw flames erupting in cities ensnared in race riots. We were not immune to all these changes, but we were not part of the vortex that seemed to be tearing apart America.

Crazy Days of Summer

It may be during my college years that the seed of horticulture was planted, but it certainly was not because of the classes I took. I believe there was a general class in horticulture, but it was perceived to be far less important than woodlot management. I decided to major in botany because it was the least uninteresting offering, and some of the subjects, such as plant physiology and morphology, caught my attention. About the same time, a new science called ecology was just emerging. I was quite interested.

Just outside the ecology professor's office was a small lean-to green-house, where houseplants in pots and baskets were grown. It was usually locked, but, when open, was one of my favorite places to visit. I must confess that many a cutting found its way into my pocket for subsequent rooting in a water glass. I expect that explained the lock on the greenhouse door. However, as far as learning anything about ornamental horticulture, not a single petunia was ever mentioned. No, the seeds of horticulture evolved away from campus, during times not in school.

During those years, I had a number of summer jobs that involved horticulture, mainly because those jobs were there. Today, students take cruises or go to the beach, while the smart ones look for internships to gain experience in their fields. My summer jobs could never be defined as internships; I took what I could find—most involved menial labor and very little pay.

It was on one of these jobs that I learned way more than I wanted to. A nursery sales recruiter showed up at a classroom at school. I listened as he told us how we could make thousands of dollars selling his company's products on the road. He regaled us with stories of his wealthy colleagues, who, like us, all started their careers as summer salesmen. Starry-eyed, I listened as he explained how we were not selling a product, but rather providing an opportunity for people to buy something they needed. More importantly, we would be paid a healthy commission by doing this good work. "Go out and make people happy" was the mantra, and we bought into it.

The job entailed selling fruit trees and plants of small fruits such as strawberries and blueberries to the people of rural Quebec. Armed with catalogs, photos of mature apple trees, luscious strawberries, and gooseberries, I jumped into my old Volkswagen Beetle in early May. I remembered my lessons and traveled the province to make people happy. After all, who did not need fresh apples or fruit on their table? It turns out I may have been a wee bit naïve.

Did I mention that rural Quebec means farmers? Or that, in rural Quebec, planting and field preparation take place in May? And that

I had to do this in French? Not only did they have no time to talk to some Anglophone who could hardly converse, they had not the slightest interest in planting apples or strawberries when every one of them had large food gardens. Being a bright student, I recalled the super salesman telling us to talk to the wife. "The women love this stuff." If I ever did reach the wife, she was feeding chickens, tending to the house, or preparing meals. If, by chance, someone showed some interest and listened to my pidgin French, things really fell apart. Did I mention that the trees and plants would not be delivered until next spring, but payment was due immediately?

The lessons in the classroom were not translating well to lessons in the field. Suffice it to say, I learned a great deal about naïveté and taking advantage of head-in-the-clouds kids, but mostly I learned about myself. As bad as this experience was, it toughened me up.

I could never be a door-to-door salesman, but I learned rejection is not the end of the world. I learned to do my homework, prepare for any job, and essentially do everything I could to reduce stress. While I would not subject a young person to what I went through, I strongly recommended to every student I have ever taught to gain some experience in sales. The skills I picked up in dealing with people, looking them in the eyes, and forging a relationship are lessons young people can take into any future endeavor. Unfortunately, it didn't do a whole lot for my love of horticulture.

I don't recall how long I lasted in that summer job—probably six weeks or so. Soon, I found another job at a small greenhouse that came with a rustic shack in which I could live. I learned how to take cuttings of honeysuckle and dogwood and a little about rooting hormones. Outside the door of my little cabin was a large bed of bearded iris, totally overgrown with grass. My boss asked me to clean it up, and I quickly learned that grass does not pull easily, especially when embedded with iris roots. I finally decided that the only way to do this job was to dig every thing out, remove the grass, and replant the iris.

This took me at least a week of after-work toil, but I mention this because I recall the satisfaction of doing the job right far more than I recall much about rooting honeysuckle. However, as far as horticulture being a career choice, it was barely on the radar. If anything, it was a lot of work for little return.

Perhaps There Is Something to This Plant Stuff After All

I had many summer jobs ranging from haying on French farms in high school to applying herbicide to milkweed and thistle in soybean and cornfields in Ontario. However, whenever I am asked, "How did you get into horticulture?" an event during my grave-digging days stands out.

I had found a job in a Montreal cemetery digging graves—literally digging with spades and shovels. I quickly learned that a grave hole is large. The cemetery advertised "perpetual care" to families, which included far more than cutting the grass. It meant that we literally planted the dirt on the grave site with all manner of bedding plants. A small greenhouse on site grew all sorts of green, gray, yellow, and purple plants—all less than six inches tall. Think of a floral clock or flag, and you can understand what we did. I knew little and thought even less about plants, but I was occasionally involved in planting them on the fresh grave site. I dug, planted, and watered these sites—but mostly dug. I should have paid more attention, but I was so exhausted after digging all day, I did not learn anything about sedums, artemisia, alternanthera, and lobelias until many years later. I did not even learn their names. Just call me a teenager.

. . . .

One day around five o'clock, I was leaning against a tree near a newly planted grave site. As tired as I was, even I appreciated how beautiful it was. As I was about to leave, a family approached the site. I slid behind the tree. They were in mourning, paying respects, and crying softly. As I looked out from my hiding place, I perceived a subtle transformation. They noticed the beauty of the planting, and their spirits lifted. I watched the despair of the moment give way to the appreciation of the beauty before them. It was almost like the dark clouds slightly parted, allowing the sun's rays to emerge. I watched with surprise that slowly turned to amazement. Neurons must have crackled somewhere in my brain. I realized that there may be something to this plant stuff after all. Caring truly made a difference.

. . . .

Such reverence was quickly squirreled away into the far reaches of said brain, not to be recalled for years. I returned to school with a sore back and calloused hands, and with no idea what I would do with my life. Horticulture was still on the far back burner.

The times at Mac were some of the best of my young life. It was a time of summer jobs, winter schooling, hitting the books, and skipping classes to play shinny hockey. I was a scholar-athlete, drank a little beer, and edited the school newspaper. My friends and I would often sit around and promise ourselves we would travel the world. I was young and not especially bright, but slowly I was gaining a bit of maturity. Like so many others of the Sixties, I lived the verse, "To Everything There Is a Season," and realized that maybe—just maybe—there is "An appointed time for everything." However, the time to embrace horticulture had not yet been appointed.

Sad Times and Good Times

My parents split up in 1964. Mom had started a job as a social worker when I was in high school and moved to an apartment in Montreal. It was a difficult time, but she was an amazing woman and was surrounded by friends and young people she had helped in the past. She was loved and stayed busy. My dad, on the other hand, was not faring as well. He was still smoking and drinking too much. When I visited his small apartment, it was obvious he was a lost soul. It was a truly sad time for him and me as well. As lousy as they were together, they loved their sons very much. They tried the best they could. On September 9, 1966, Dad had a major heart attack and died that evening. He was fifty-five years old.

My brother was also having his share of problems. Having done a few ill-advised things while in college, it was not apparent that he would even finish school. While taking a break from college in Montreal, Hasie did the smartest thing he has ever done. In 1964, he married his longtime girlfriend Phyllis. From the day they were married, Phyllis turned him around. Together, their accomplishments have been astonishing. Maybe this was a sign of good things to come.

I met many people at MacDonald College and have many fond memories. One of those memories was a class ski trip to Owl's Head Mountain in the Eastern Townships of Quebec. Having seldom gone to the mountains as a kid, I was a poor skier. I was also discouraged at every turn by Coach Baker, who understandably worried about knees and other body parts being injured. However, I was tired of staying home weekends, so off I went to the hills.

Love at Owl's Head

Once I figured out how to stand up and not crash into a tree, I noticed a beautiful girl with a wonderful smile. I inquired as to her name: Susan Downman. She ignored me completely, which was not unusual, but I watched her intently. She flew down the hill, and effortlessly floated down the slopes. Susan epitomized beauty, grace, and athleticism. I was hooked.

I made it a point to meet her. I soon discovered a person admired by everyone for her character, wit, and intellect. She was everything I was not.

Susan Downman, 1967

Unfortunately, Susan wanted nothing to do with me. But I was persistent, and she finally relented. She agreed to join me at a movie and confessed to ignoring me because I was irresponsible and foolish. Of course, she was correct. We laughed, we played, we even tried studying together—once only. She was not at all impressed with my study skills. When we graduated, I wasted no time in asking her to marry me.

She has been the rock of my existence ever since. She has guided me every step of the way, and without her, there is absolutely no doubt I would not be where I am today. It is quite likely I would be flipping burgers at the Orange Julep® in Montreal.

I had learned a good deal about myself in the four years at Mac, but still had little traction concerning my future. I was twenty years old. When I was told to move on from college, the next eighteen months saw changes I could not have foreseen.

Four

Vegetable Juice and Mushroom Soup

I had steadily brought my grades up since year one, and in spring of 1967, I graduated with gusto. I had earned a very respectable overall percentage of seventy-three percent. One of the unforeseen consequences of being a reasonably bright student came much later when students like myself applied to American colleges. A grade of seventy-three percent would not even get you past the registrar's secretary. But that is for another day.

Upon graduation, I accepted a job with Canadian Canners, a division of Del Monte Foods®, in Hamilton, Ontario. Canadian Canners was a food processing company. They canned vegetables and other food products under the brand of Aylmer Foods®, a well-known label in Canada.

It was a dream job. I was paid $6,500 a year—a veritable gold mine. It may even have had benefits. Who knew what benefits were back then, and what young kid cared? I recall buying my first ever suits—two of them, fitted no less—each costing the unheard of sum of $100.

My makeover only extended so far. Confident as a peacock, I walked into the office wearing my charcoal-gray suit and white socks. I really was a doofus back then. Can you imagine a new suit—and white socks?

I was told that I was on the "management" track and asked to perform a number of different jobs. One was Quality Control Supervisor, whose responsibility was to check the quality of vegetables being canned at various plants across southwestern Ontario. I was too new to realize that a visit from the home office was the last thing a plant manager wanted—from a young whippet fresh out of school even less. During the visits, I walked the facility, stopped at each line—perhaps beans or peas or tomatoes, depending on season, and inspected them for disease, discoloration, size, etc. In short, I was checking up on them, and while they were pleasant, they were quite happy to see me leave.

Plant managers were paid for efficiency. The more cans that ran through the lines, the better their paychecks. Anything that slowed down the line, such as the truck carrying beans being late, hurt their efficiency rating. And heaven help the person who did something to cause the line to shut down for any reason. While I was aware of this, my youthful enthusiasm may have gotten the better of me.

I knew when I walked into Plant 41 that there were some problems with the corn. There were too many broken kernels, too much bruising, and too much debris from the load. I asked the manager, a veteran of twenty years, to do a better job of eliminating such problems prior to the product being put on the canning lines. However, no improvements were obvious ten minutes later. The manager said it was fine, I said it wasn't. I told him to shut down the corn lines until the problems could be repaired. To say that he was irate is an understatement, but after a rather intense discussion, the lines were stopped. More people were then assigned to cull the offending kernels and detritus, and the problem was fixed. The encounter was memorable and not at all pleasant. Back at the office, I was told that significant inventory was lost. In the future, I was to shut down the lines only in the most egregious situations. I reflected a wee bit on whether this was my dream job after all....

Things dramatically improved when I was asked to assist in the New Product Research Department, where new products or improvements to existing products were explored. Like mad scientists, we added a cup of this, a dash of that, or a dollop of something else, cooked them up, and then compared them to products already on the market. My first project was to make a vegetable juice to compete with V-8® juice, by far the leader in the market. More carrots, fewer beans, more cilantro, less salt, more oregano, less celery, and on and on. After each batch, we would taste it. If a particularly tasty batch emerged, we would compare it with V-8.®

The staff employees were our guinea pigs. When we announced a new batch was ready to be tested, it was amazing how many people were away from their desks. If we had a winner, the recipe would be secreted away, and the line would be tweaked. Changing a recipe was like changing a car model—requiring line stoppage, significant retooling, and a new advertising campaign. With so much at stake, we had to be sure our new recipe was significantly better than what we already had and equal to or better than our competitor. We did improve our vegetable juice, sales rose, and management was pleased.

My next project was to concoct a better cream of mushroom soup, one to compete with Campbell's®. Similarly, we shifted this, augmented that, added more milk, more mushrooms, different mushrooms, or different seasonings. We sampled and tasted one batch after another. Over a few months, we created a better mousetrap. I have not had cream of mushroom soup since.

I very much enjoyed the research, and there was little doubt that I was on a fast track for an administrative position. In a few years, I would be making very good money and enjoying my own office. Sometimes I look back on those days, and wonder, "What if?" I was doing well, climbing the corporate ladder and living the Canadian dream. But part of me simply was not there. A little gnat in my brain kept saying there must be more. I remembered the times as students, sipping beer and vowing to ride motorcycles through Europe. It was the 1960s. It was a time to explore, a time to see the world—and there I was making soup. Something was missing, and I knew what it was. My next decision was ill timed, ill conceived and quite ludicrous, but I had to do it.

The Wanderlust

I had been in school my entire life, worked during summers, all to compete for a job. Without realizing it, those years quickly passed, and here I was, working again, this time likely for the rest of my days. I was restless, simple as that, and not ready to settle down. I wanted to get away, travel, and see more of the world. Most of all, I wanted to do it by myself. As to the latter, that was no problem. None of my friends was going to pick up and go halfway around the world on some half-baked voyage. However, there were a couple of obvious flaws in my plan.

First, I was engaged to Susan. She was working as a dietetic intern at the Royal Victoria Hospital in Montreal, while I was seeking my fortune in Hamilton. We would visit, write, call, and pine for one another. When her internship was completed, we were to be married. When I told her my travel plans, she was not a happy lady. I can imagine thoughts of her fiancé traveling through Europe, getting into all sorts of escapades with wine, women, and Lord knows what else coursed through her mind.

If she had yelled, screamed, carried on, or said no, I would not have left. There was no good reason to go except that it was an itch needing to be scratched, and I would not have risked my future soul mate. We had a few discussions to be sure, but one day she simply said, "If I say no, you will resent me, and that is not a good way to start a marriage. Of course, I don't want you to go, but I think you should get it out of your system, so that one day, I won't wake up and find you gone." That was only one of the dozens of times Susan's perception and wisdom came to the surface. And so, I quit my job.

When I speak to young people today, I often quote the baseball legend Yogi Berra. I find many of his sayings quite profound. One I have always felt particularly appropriate is, "When the path forks, take it."

Here was a path forking. It was one of the important paths for both Susan and me, and we decided to take it.

There was one other flaw in my grand plan. I had no money to fly overseas, and certainly not enough to spend five to six months there. I had heard that merchant ships occasionally took people on board to offset expenses. I don't recall how I discovered this, but I looked into it, and, sure enough, there were steamers that crossed the Atlantic offering inexpensive passage. The rules were simple. Stay out of the way, and get off at the first port of call. In mid-March, I took a bus from Montreal and arrived at the docks of Saint John, New Brunswick. I paid a passage fee of $155. The next morning, I jumped on an old tramp steamer. I made my way down decks to a converted storage area that was to be my living space for the voyage. I expected few comforts, and that is what I received. It had a light, a bed, a desk, a porthole, and not much else. But for me, it was just fine. It was a means to an end—I did not expect a luxury cruise. That morning, I steamed from my beloved Canadian soil into the cold, wintry North Atlantic Ocean. First port of call: Dublin, Ireland.

The captain was understandably wary when I told him this was my first trans-Atlantic crossing. He proceeded to talk about life jackets, storms, ropes, waves, cold, and safety in general. I understood every third word of his heavy Irish accent, but I came away with an appreciation of the dangers of the crossing and the unpredictability of the ocean, particularly at that time of year. I also came away knowing where the life jackets were stored. I was certainly an oddball to the mostly Irish crew. A passenger was a new thing for them. While they were a bit curious, we coexisted well. My days were mine to do as I liked. I spent most of my time reading, walking, and taking the occasional nap. I was introduced to the traditional Irish breakfast of eggs, sausage, boiled tomatoes, mushrooms, and dry toast. Lunch and evening meals consisted mainly of meat and potatoes. I had my first taste of Guinness® stout and rather enjoyed it. The pace of life for the crew on the ship appeared to be routine, and mainly consisted of maintaining the equipment and constantly painting areas of the deck.

Today, in the time of cruise ships and inexpensive group flights, the practice of merchant vessels taking on a few passengers no longer occurs. Even then, it had to be a very limited profit. Liability laws today would never allow for inexperienced passengers to roam freely about a ship where the only thing between the deck and the ocean below was a four-feet railing. Still being in my young and stupid phase, I saw no problem. I can't imagine the captain was at all pleased to have such a responsibility added to his tasks. Truly, when I reflect on this voyage, I think I must be making this up. But there I was, bobbing along the ocean in my cubicle, or more often, walking around the decks, leaning on a rail and staying out of the way. I was just getting into the routine of ship life when day three blew in.

. . . .

I awoke clinging to my bed as the room pitched and heaved as if in a salt shaker. We had entered a North Atlantic storm. The crew was roping stairwells and the deck so they could move about safely. I proceeded up the stairwell, pulling hand over hand on the rope, and eventually reaching the deck. The sky was gray and forbidding. The wind cold and furious, and the water white and menacing. The ship was tossed about like a yo-yo. The crew was unfazed. This was not a violent tempest. I remember someone yelling at me as the storm intensified. "Walking around clinging to ropes is not a very bright endeavor" in words a great deal more descriptive and salty. However, when was I ever going to experience a North Atlantic storm again? I clung, balanced, shuffled, and bobbed from rope to rope. I braced myself against the wind and spray for all of fifteen minutes before sense won out over machismo. I went below deck and had a bit of food, then returned to the relative safety of my room. As predicted, the tempest grew worse. I could stand the heaving and the lurching, but being stuck beneath decks during a storm was claustrophobic. I had to go up every now and then to breathe the stormy air or I would go mad. I would make a lousy prisoner. Fortunately, I was not prone to seasickness. While the day was long, it was an adventure I would likely not confront again.

. . . .

It turned out it was not a particularly violent storm and we were never in any real danger, but it seemed violent enough to me. As it turned out, my turns about the deck did not go unnoticed by the captain.

The next morning he complimented me on my sea legs and asked how I was enjoying the trip. I stated all was well but, March storms notwithstanding, I was a little bored. He wondered if I would like to do some work around the ship. Without waiting for my reply, he handed me a tin of paint and brush and pointed out some darkened rusted parts of the deck, and away I went. I was ecstatic to help out. I painted only a few hours a day, but felt somewhat useful. On top of all that, I got my $155 back. On March 21, 1968, after nine days on the ocean, I was deposited on the dreary docks of Dublin, Ireland.

Footloose and Fancy Free

It was the beginning of a six-month journey through Ireland, England, and the European continent. I was on my own, with no transportation, little money, and no means of earning any. I looked about me, took my bearings, found a road into the city, and stuck out my thumb. Before long, I was picked up, deposited in Dublin, and found an inexpensive bed and breakfast. Thus did I learn the meaning of self-reliance and self-determination—not to mention a passion for travel and an appreciation of history. My thumb took me around all of Ireland then to England. I stayed in London for nearly a month, then traveled north through France, Germany, Belgium, and Sweden, and as far south as the Greek Islands.

Dozens of adventures awaited me; interesting people befriended me. Cities and towns I had only read about in history books left me speechless. When I arrived in London after three weeks in Ireland, I happened upon a rather rundown bed and breakfast in the Victoria area. Paddy, a young Cockney fellow, and his mother ran it. To raise money for accommodation and to have money for food and drink, I had to find work. Paddy painted the exteriors of apartment buildings and

Allan in France, 1968

asked if I wanted to join him. Since I had so recently become such an expert painter, I agreed readily. These were not skyscrapers, but old buildings up to five stories high in southwest London. I learned ladder craft: the ability to walk with a fully extended twenty-four-feet ladder from one spot to the next. I also quickly learned how to handle a rope supported suspended platform. Paddy was beyond cheap. He rented the type of platforms that were secured to the roof with cement blocks or any other kind of ballast. The platform barely held the two of us. In order to raise or lower it, a system of rope pulleys was used. Releasing some rope lowered one side, and squeezing the rope would stop the descent through friction. We repeated the steps on each side of the platform until we achieved a semblance of being level. If OSHA's equivalent existed in London then, we would have been shut down before we started. But we were young and invincible, and most of the money I earned was saved for tomorrow, while a tiny bit went to the local pub—well, maybe more the other way around.

For six months, I traveled and worked. I painted in London, tended bar in Geneva, did "Little Joe" card tricks for a few francs in Germany, and tasted beers across Northern Europe. I sipped red wines in Italy, ate meatballs and loganberry jam in Sweden, Wiener Schnitzel in Bavaria, baguettes and cheese in France, and souvlaki in Greece. I saw sunsets in Skagen in Denmark, awoke to sunrises in Biarritz in France, inhaled my first sample of weed in Italy, and slept on a train through Yugoslavia. I grew my hair long in Denmark, wore sandals in France, and carried a hand-woven shoulder bag in Greece. In short, I did what I had to do to see what I didn't even know existed. Not a single thing did I learn about horticulture, but I had become educated all over again.

Later, when I taught high school and then at the university, I looked at my students and emphasized the importance of travel to every one of them. There are few things more important in one's education.

Susan would not let me go on this trip without a return air ticket in my pocket. In mid-September, I was on an Air Canada flight from Athens, Greece, to Montreal. She tells the story about how she was waiting in the airport and spotted me from the back, looking handsome and debonair in my new European clothes. As she approached, arms out and smiling, she felt a tap on her shoulder and turned around to see— me—long hair, sandals, and a purse. Her smile faded, and she shrieked a little, but the reunion was wonderful. That's her story, and she is sticking to it.

Get Me to the Church on Time

I met many young people who also embraced the lack of responsibilities and the freedom that such wanderlust allows. I also met many people who savored the lifestyle and stayed too long. I loved every aspect of my adventures in Europe, but six months was enough. I would return some day and take Susan with me, but for now, there was a whole life waiting. I knew what I was going to do. First, I was going to marry my incredibly patient fiancée, and then I was going back to school to become a teacher.

We wanted to marry right away. Susan's sister, Sandi, was married in June

Allan back from Europe.
Notice the the spiffy footwear..

The Happy Couple, 1968

of that year, which sapped the energy from Susan's parents. While we did not want a lot of fuss or bother, they kept telling us the next summer would be best for them. Although we were both in Montreal, this commuting relationship was quickly getting old. Susan's family was very "British." Her grandparents still had afternoon tea (as did mine), and Victorian mores were alive and well in 1968. After nearly going off the road one night returning to my apartment, we decided it was time. We told them that if you do not allow us to marry this year, we would move in together. This is the norm today, but, oh my, shock and awe accompanied that statement in the fall of 1968. The ultimatum was granted with the belated blessing of all, and we set our date for December 21 for family and a few friends. My best man was to be my brother, who lived in Ontario with Phyllis. Wouldn't you know it—he came down with the Hong Kong flu and was unable to attend. A big snowstorm added to the excitement; I skidded out on the way to church and got stuck in a snow bank. The song "Get Me to the Church on Time" was ringing in my ears as I madly tried to extract the car from the snow bank while in my black shoes and best suit. Soaking wet and freezing through and through,

I finally staggered into the church to the relieved sighs of family and friends. We were twenty-two years old, young by today's standards, but through thick and thin, she has stayed by my side.

I believe the total number of wedding attendees was about twenty-five. The total amount of money spent was miniscule, but a grand time was had by all. When I attend some of these incredibly lavish, extravagant weddings today, I ask myself, "What is the point?" Too much money, too much stress, too much worry—but what do I know? I wonder if studies have been done to correlate quality (and length) of marriages with money spent.

Learning How Not to Teach Teachers

I was ready to settle down. Somewhere along the way, I believed I might make a decent teacher. I am not sure if the classroom enamored me or if it was a relatively easy path to follow. Upon returning from Europe, I immediately enrolled in a one-year teaching certificate at MacDonald College, and the newlyweds took up residence in married student housing, affectionately known as the Huts. Such a moniker was a sizable embellishment, they had four walls, a roof, some heat, and no phone—and in the winter of 1969, were buried in a ton of snow. While I was living the life of comparative leisure, Susan was commuting twenty miles to the Royal Victoria Hospital in Montreal, where she worked as a special needs dietician.

Driving in a Montreal winter is demanding at best, and quite dangerous after dark, especially when it snows. She left before sunrise and returned after sunset, and it snowed a lot. As inexpensive as the Huts were, her income was essential. And, as bad as the accommodations were, everyone there was in the same situation: we were all broke, and all cold.

Allan and Susan in Front of The Huts

Regardless, lots of laughs and friendships developed.

. . . .

In the winter of 1968, Susan developed a severe case of the Hong Kong flu. She was so ill I was scared that she was not going to make it. Every part of her body ached, including her hair and teeth. She could not retain any fluids or food and was essentially comatose. Everyone seems to complain of having the flu when they experience a bit of coughing and fever, but they have no idea. It is not until you have truly experienced "the authentic flu" that you understand why so many people have died from it. She slowly recovered. Life returned to our broke, cold routine.

. . . .

I took classes on how to become a high school teacher. Today I read about incompetent teachers, incompetent students, and incompetent administrators—none of these are new developments. As for me, it was a rather incompetent curriculum. I recall very little of the classes, but I was focused and ready to learn. I remember the importance of learning audio-visual skills, like threading the movie projector and using an overhead projector. I am sure there were also suggestions about discipline and respect for your principal and handling parental complaints. But there was nothing about the joy or passion of teaching. Nothing about the ability to transform lives of young people or watching the light go on in a math class when a student "finally gets it!" These were all things I experienced when I got into the classroom.

What I do remember from school is filling out reams of paper to make daily lesson plans. What a total waste of time it all seemed. If one were not sure if one wanted to be a teacher, this program would have drained all energy and confidence within weeks. Finally, we were assigned a school to do practice teaching, first for two weeks, followed by a six-week stint in the classroom. By the third day, I knew that this was what I wanted to do.

Being in the classroom was the only thing useful about the teaching program. I suppose that in itself made the year worthwhile. That, and the fact I was still eligible to play one more year of college basketball. The saying, "I may be slow but I'm old" was apropos of me. I managed to do just fine, the team did very well, and I connected even more with Coach Baker. So much so that he offered me the post of assistant coach the following year. I remained in college coaching for six years.

Wonderfully Good and Incredibly Sad Times

We resided in a small apartment in Lachine, Quebec. When spring finally arrived, that horticultural seed must have awoken also. I decided that we should grow some flowers in a window box hung from our third-floor balcony. The box took up the entire length of the railing— I worried the building might topple under its weight. Susan called it the "coffin." I planted seeds of marigolds, bought a few petunias, and forced some begonia tubers in a small lean-to greenhouse associated with the school. The garden germ was starting to coalesce. Then life changed.

In 1971, Susan was working as a dietician at one of Montreal's hospitals. I was teaching, and life was good. Although we were both healthy, Susan started getting violently ill in the morning. It became obvious that she was either coming down with severe food poisoning, or she was pregnant. A trip to the doctor and her body shape soon confirmed the latter. We were ecstatic. Tales of ice cream and pickles, mood changes, and fatigue all proved true. Each visit to the doctor confirmed all was well, and she worked until the eighth month. How well we still remember her discomfort, how she wanted this little person to appear, and how much we wanted to bring a baby home. Our daughter Shannon was born on June 18, 1971, and died twenty-four hours later.

Shannon was born with a hole in her lung. No warnings showed up during any pre-natal visit. At her birth, there was a tension in the air as the doctors and nurses realized something was not right, but I don't recall bells going off, emergency people gathering, or anything. Susan was able to hold her for a little while before she was taken away, and that was the last we ever saw of her. Her funeral was a week later, and she was buried in the Mount Royal cemetery.

No one can prepare you for a loss of a child. No one can understand what a mother goes through when such a tragedy occurs. Friends and family were supportive, but Susan experienced major depression. It is not so much the initial tragedy that consumed us—everyone was stunned, and arrangements kept us busy. It was the weeks and months after her death that were the hardest. Life seems to stop; there is no understanding of the whys and wherefores, and, although people around us did their best, we were alone within a black cloud of despair. I remember Susan remarking that some of her hardest times were not reliving it with friends who came to comfort her, but times when friends were so afraid of upsetting her that they never mentioned the baby. She remarked that if ever she met a mother under similar circumstances, she would go out of her way to talk to her. Pretending it did not occur or speaking about everything but her loss was far worse than talking about what happened. Unfortunately, friends and colleagues have experienced similar losses, and Susan has been there to comfort them. A number of years later, when Susan went back to nursing school, she became a labor and delivery nurse.

Getting back on our feet was not easy, and those who say, "Get over it," never experienced "it." However, we did resume living and were more certain than ever that we wanted a family. We bought an old Volkswagen camper van and during late summer of 1971, we drove to Nova Scotia and Prince Edward Island. Volkswagen campers were wonderful things, providing wheels for travel, a place to relax at a campground and a cozy bedroom. Suffice it to say, we took full advantage of all the features of our camper, and, upon our return, Susan was once again having trouble retaining her breakfast. A healthy Laura Armitage entered the world on May 19, 1972.

Dr. Spock We Were Not

All new parents are a little shell-shocked by the arrival of a baby. They are proud and believe their child will be the smartest and cutest, and that they love this child more than any parents could ever love a baby. By the time the second or third one comes along, not so much. But with the eldest, how we can mess them up with love, attention, and Dr. Spock. We were no different. Every new step, new word, and smile was cause for pride. Every misplaced belch, temperature spike, or new tooth was cause for consulting Dr. Spock. We believed she was the cutest and brightest baby ever born, but we could not believe the length of this newborn.

. . . .

I could not believe I was in the delivery room watching Susan give birth. No one ever told me there would be so much groaning, grunting, pushing, and panting. Although I certainly wasn't ready for the decidedly unladylike position she was forced to endure, Lamaze had trained me to help her during delivery. As we were puffing and pushing together, I heard the doctor say, "That's good, keep pushing, the baby is coming. Yes, she's coming, keep pushing, oh my God, she's still coming—STOP PUSHING!" When she finally tumbled out into the world, Laura was measured in at twenty-four inches long. I do believe she was the longest, skinniest kid that the doctor and nurses had ever seen. My daughter, Laura Elizabeth, was two feet tall at birth! Good grief, if a child is two feet tall at birth, how tall would she be when she stopped growing? When she was carefully lain down on Susan's stomach, she didn't fit. When I picked her up, she kind of buckled at either end and formed a U-shape. It's tough to cuddle a kid who is almost as tall as you are.

. . . .

We left the apartment and the plant coffin behind three years later. We found a beat-up bungalow backing on a highway, but in the lovely neighborhood of Pointe Claire, on the West Island of Montreal. As soon as we'd settled, I wanted to do a little gardening. I noted we had some variegated dogwoods, some half-dead junipers, a few yews, a gooseberry bush, and a ton of lilies-of-the-valley, otherwise known as Canadian kudzu. In other words, we had the same plants seen in every Montreal garden. Like most males, my first love was turf. I smothered that bluegrass with fertilizers, poisoned it with weed killer, and drowned it with water. Perhaps because throwing things on the turf was brainless, perhaps because there was more turf than anything else, or perhaps because my competitive genes were still alive and well, I was the epitome of a television commercial for You-Name-It Fertilizer Company. It was not a pleasant sight.

I worked on some beds around the house. Lo and behold, I discovered that I liked digging in the dirt. I ordered hyacinths, daffodils, and tulips and studied up on roses and peonies. I believed all the outlandish descriptions of exotic plants in catalogs. I marveled at the photos, knowing full well my plants would be equal to any pictures in a catalog.

Yes, I was a gardener! My desire to plant was far greater than my knowledge—I killed my fair share, but also managed to succeed with a few. Between the turf and the additional shrubs, perennials, annuals, and bulbs, the yard slowly became a garden.

Five

"A teacher affects eternity;
he can never tell where his influence stops."

. . . .

- Henry Adams

Good Morning, Mr. Armitage

I accepted a position of biology teacher in the Science Department at Beaconsfield High School on the West Island of Montreal, teaching two different level classes to junior and seniors. I also taught math to middle-school-aged students. Canada has no middle school designation, so young and older kids are housed in the same school. Public school teaching then was quite different from teaching today. Not any harder or easier to be sure, but there were far fewer academic rules and less teaching to the test.

We certainly had more ability to talk with kids or hug them without fear of being labeled a pedophile. Teaching is one of the most rewarding of jobs. Thousands of teachers make an incredibly positive difference in the lives of children. It is a calling more than a job. You must learn to live with the chaos of too many classes, too many hormones, too much grading, and, unfortunately, too little money. It was through high school teaching that I learned skills of communication. It paid me back in spades in the university classroom and on horticultural stages across the world.

I have seen my share of poor teachers; a few because they never should have been there—but in what other profession does that not occur? If they did not leave after a year or two, they simply took up space. I met many teachers who were proud of teaching twenty years, but, in reality, they taught one year twenty times. Others were simply worn down over time. But to be clear, I met many extraordinary teachers. Public school teaching is one of the most difficult callings in which to excel. Day in and day out, a good teacher must be on stage—teaching not only the subject but also the entire student—six classes a day, each day, every week, every year, with enthusiasm. Rewarding as it may be, it is impossible not to wear out a little.

I loved teaching at the high school level. I even managed to instill a little knowledge of biology and math to the hundreds of faces in my classes. I was evolving, if only a tiny bit, into more than just a teacher. I was a good listener. Sometimes that is all that troubled kids needed. In addition to coaching at MacDonald College after school, I volunteered to coach volleyball and woman's basketball at the high school as well. In order to become a good coach, I just thought about my high school basketball coach. I simply had to do the opposite—and I did just fine.

Outing with Mr. Armitage

I vividly recall two things that enhanced my experiences as a teacher. One was the Outers Club. Not everyone cared about sports or the debating team. Extracurricular activities depended on the staff's willingness to give significant additional time—for no remuneration. Teachers were tired of students (and *vice versa*) after days and weeks in the classroom, and most simply didn't have the desire or time to do more. I was still young and energetic, so I started an outdoor club to see if anyone was interested in things such as camping or canoeing. A few students actually showed up. We formed the Outers Club and did some hiking in nearby woods to appreciate the trees and wildflowers. We were laid back and enjoyed each other's company. Soon more kids showed up, and we planned some activities on weekends, when we would go out a little further. We talked about possible outdoor activities, then went out and did them. I was the leader, the representative of the school, and the protector of the children. No administrator or parent or school board member sanctioned, approved, or even read over what we were doing.

Looking back from where I sit today, a number of things are self-evident. One, I was young and never considered that any misfortune could occur. Two, I had no training in most of the activities I was leading. It is amazing that no one got seriously injured. And three, this would never happen today. We went whitewater canoeing (I'd never been whitewater anything), we camped all over the place (I knew nothing about camping), and we even did some rock climbing (ditto). We scrounged, we borrowed, and we traded for tents, ropes, canoe rentals and more. We simply faced the problems and tried to figure out ways to solve them. In general, most outings were challenging but not overly daring; however, two experiences could have been disastrous.

One of the students, Charles, had done some rock climbing with his dad. He had ropes and equipment we could use for climbing and rappelling. This seemed like a good idea, so off we went to some serious cliffs a few miles away. Keep in mind that these were suburban kids who had probably never climbed a rock face and certainly had never rappelled down one before. The leader, me, was not happy about falling off a log, let alone a cliff. Despite my utter incompetence, I led them up the cliff. Charles provided instructions about belays, knots, and carabiners, and how to fall off the peak to begin the descent. The instruction lasted all of a half hour, and away we went. I will spare you the rough details, but suffice it to say if you want to build confidence in yourself and in your team, fall off a cliff. Do so while knowing that teenagers holding a rope are the only things between you and the bottom of the cliff. Amazingly, we all did it. Even more amazingly, I never worried about the potential dangers. Never a fleeting thought that maybe a fifteen-year-old instructor and a young teacher with no experience being in charge of the safety of ten children on a windy cliff was not really a great idea. In our blissful ignorance, we repeated it a number of times. The kids became so confident they even wanted to rappel from the school roof. The administration did not think that to be a good idea.

I remind you that we were doing outdoor activities during the school year in Montreal, and much of the school year in that part of the world is winter—and cold. Somewhere I read about the thrill of winter camping. At one of our after-school meetings, I brought in some articles about building a snow shelter, how to use green wood to reflect the heat from the fire, finding dry wood for said fire, the benefits of paraffin-dipped newspaper rolls for long lasting candles, and proper food for eating around the campfire. We discussed responsibilities, warm clothing, and proper sleeping bags. We realized it would be cold enough that liquids, like water, would likely freeze. We brought pots for soup, hot chocolate mix, and hot dogs for easy grilling. After all, it was only for one night. Gourmet meals were not on the menu.

About ten of us went off on a beautiful Saturday morning to a property in the country owned by a family of one of the students. It was already a snowy winter. We managed to find snowshoes for all; otherwise, we would sink waist deep in the snow. We walked about a half mile to the site to set up camp. The sun reflected off the snow, the temperature was a balmy twenty degrees Fahrenheit, and soon we were sweating in our parkas as we built a snow shelter. Some of the students lined the fire pit with green logs; others searched for kindling and dry wood. We all clomped around the camp in snowshoes packing down the snow. Some of the kids even built two- or three-person mini snow shelters. We seemed to be on top of things.

Not far from the campsite was a log cabin, our Plan B if things did not work out. No one had any intention of going there; this was an adventure.

About three p.m., the sun began retreating behind the fir trees. Temperatures were rapidly falling. By five o'clock it was dark and getting colder. The fire was blazing and our paraffin-soaked newspaper rolls made surprisingly good lanterns. We sat on dry cardboard, talked, sang, and laughed. By six, the temperature continued to plunge. We decided it was time to eat dinner. We had poured water into a number of pots when we first arrived, and of course all were frozen solid.

We placed a pot on the fire for hot chocolate. The water eventually came to a boil, but by the time the cups were passed around, the hot chocolate was cold. The soup was so frozen we could not dislodge it from the cans. The moment of, "Oh geez, this may be colder than I thought" arrived when we could not get a skewer into the hot dogs. Every wiener was as solid as steel. We tried metal forks, sharp twigs—anything we felt might work. The buns simply fell apart like confetti. Marshmallows had become cubes of granite in the freezing cold air. We threw the hot dogs on the sheet of ice in another pot but no matter how hot the fire, cooking ten hot dogs in ice was simply not feasible. We decided to keep the remaining water for drinking if needed. Although we were surprised at this turn of events, hunger was not the problem. It was getting colder.

The stars shone bright in the beautiful clear sky. There was nothing to do but shiver around the fire. We decided to climb into our cozy sleeping bags in our less-than-inviting snow shelters. I stayed by the fire to keep it going, but it provided little enough heat for me, and none for the kids. I may have been stupid, but I wasn't crazy. I had no idea how cold it was, but I decided it was too cold to be lying like mummies in snow. I called to everyone telling them we were going to the cabin. Sighs of relief filled the air. None of them wanted to be the one to cry uncle, but even the toughest of the group was happy that Mr. Armitage finally came to his senses. We slowly tromped to the cabin, lit some kerosene lamps, started a fire in the fireplace, and gradually warmed up. I took a lamp to the window and checked the thermometer; it was negative twenty-five. It had been an adventure, to be sure, but no one thought we had given up; rather we all learned that caution was indeed the better part of valor.

I was worried that some of the kids may have suffered from frostbite, but only one person did. That person was me. My less-than-sterling boots allowed air inside, and my big toe tingled for weeks. It still burns when I am in cold weather for any length of time. The kids had great stories to tell in school on Monday, but that was the last time I ever went winter camping.

Red Hats Mean Pride

Many positive things resulted from these crazy escapades. The most obvious being that as they did more together, the kids developed confidence in themselves they never knew they had. Students who had never met "had each other's backs." It was a small group of normal teens, grappling with the angst and problems teens have always experienced, but the personal skills learned together reduced some of those issues. None of these results was planned, but what a team these kids became. They became so confident that our next endeavor actually made sense.

School cafeterias, by definition, are a mess. Whether food has been spattered, drinks spilled, or junk simply left around, the concept of cleaning one's table after eating does not compute to fifteen year olds. Plates with half-eaten food, cups with toxic-looking liquids still in them, tables sticky with goo, gunk, banana peels, and apple cores littered the place. The staff assigned to cafeteria duty hated it. They did their best to cajole and threaten the worst offenders, but as soon as they went to their own lunch or turned their backs, the place became a cesspool once again. It is not that there were regular food fights or overt spilling or slopping; many kids simply had no respect for the school. Custodians were expected to clean the cafeteria, which they did once the students were gone. However, to have them clean while the students were eating would cost the school additional money (combat pay, I think), would be done under duress, and would teach the kids nothing.

The kids in the Outers Club constantly bemoaned the lack of funding for rental expenses, such as canoes, vans, and tents. One day after lunching in the cafeteria, I came up with an idea. I approached the principal and proposed that the kids in the Outers Club "patrol" the cafeteria to exert peer pressure on students to clean up. I suggested that the extra money needed to pay the custodial staff be set aside for clubs like ours. The principal did not believe that kids could do any better than his staff. Nevertheless, doubt oozing from every pore, he agreed to try out the idea for a month. It was a nominal sum, but I was ready, and so were the outers kids.

Each kid donned a red baseball hat. Two or three would walk around the cafeteria asking their friends to put their trash in the can. They were provided clean rags where necessary, but for the most part did little but encourage. When asked by their friends what they were doing, they replied they were raising money for their club. Of course, they were often ignored and made fun of, but over a week's time, the place became noticeably cleaner. Kids started calling them the Red Hats. Other clubs and groups soon approached the principal and wanted to do the same. Within three weeks, while not immaculate, the cafeteria was a place students were happy to sit in. The administration was pleased, the staff was relived, and the Red Hats grew in popularity.

Two things became self-evident. The first is that confidence encourages confidence—this was true then and equally true for any-body today. Allow a person to grow, and the confidence in their abilities will only enhance those of others. The second is a bit more specific—messiness breeds messiness or its opposite—clean encourages clean. After a few months, the Red Hats were hardly needed. No one wants to sit or work in a gutter, and once the kids were used to a clean area, they kept in that way. This was by no means a model school with model students; there were still the usual academic problems, staff issues, and normal teen behavior, but writing off teens as unstable or unreliable made no sense at all.

The Outers Club was good for the kids—their parents often told me how much their children came alive when talking about it. And I was not kidding myself either; it was good for me. Perhaps it was a little over the top, but teaching at that level is not simply dissemination of information. It is offering to be involved, a little or a lot. When you sign up as a public school teacher, you sign up for the whole deal.

Of Fruit Flies and the Littlest Angel

I can't recall the number of times I had to go over and over basic algebra, rethink ways of teaching it, and then do it again and again. It is politically incorrect to suggest that some kids will simply never

Mr. Armitage, 1975.
Drawing by Jim VanHorne,
one of his students.

"get" math, but I was there. I am afraid it is so. Biology was different. In biology class, we did the usual eyeball dissection, and every year, we pricked each other's fingers to draw a drop of blood for blood typing. From the squealing and groaning, you would have thought we were amputating fingers, not poking a pinhole in one of them. Can you imagine doing such a blood-lab today? It simply would not happen without affidavits and consent forms in triplicate. We bred fruit flies to study inheritance of eye color. Under the microscope, we learned about dominant and recessive genes. There was ether to sedate them, agar to feed them, and there were damn fruit flies everywhere. Biology class was a magical place.

Christmas was always a special time. For students and staff, it was a welcome break. Schools are rather sterile places. Having young kids of my own, I wanted to finish the semester with a Christmas theme. I often cry when I see Hallmark commercials on television, so it should not be too surprising that I would do something schmaltzy. In each of my biology classes, I lit candles on the last day before Christmas break. The kids quietly entered the candlelit class. After a few minutes, I played a scratchy record of "The Littlest Angel" recorded by Loretta Young. If that was not sappy enough, I then played "Blowin' in the Wind" on my harmonica. Really, I cannot make this up! Looking back, they probably all thought I was a crazy man and surely snickered behind my back. But no one missed that class.

High school teaching was rewarding—exhausting but rewarding. I so admire longtime public school teachers at every level who can talk about their kids with passion year in and year out. You simply need to talk with them for five minutes, and you know they are special. If you have those teachers in your school, you and your children are indeed fortunate.

Mr. Lincoln's Rose

It is a shame that education is such a political football. Whether in Canada or the United States, politicians fight over control of the curriculum and budget. For politicians, there are always priorities higher than education, regardless of what they say. Whether it is road repair, social services, or pet projects, invariably the public education budget is purloined. Teachers' salaries stagnate, schools decay, class sizes rise, and lofty excuses are made. By the spring of 1974, teachers in Montreal had had enough—we voted to strike. I was as apolitical as most, but there I was marching in a picket line, holding a placard that read, "Help Teachers Teach, Not Starve" and other equally profound banners in English and French.

Unfortunately, the politicians were not moved by our eloquence. It became obvious that the strike would drag on. We were not being paid, and I needed a job. I applied to the local K-Mart and was assigned to the lawn and garden department. Thus appeared another small stepping stone in my journey. I sold lawn mowers of every type, which, at that time, consisted only of gas or manual-push types. I also talked to consumers about plants. I knew very little, but that did not stop me from suggesting this or that. My favorite group of plants became the hybrid tea roses, their roots wrapped in shiny plastic and branded "Mister Lincoln', 'Queen Elizabeth', 'Perfume Delight', and other magical names. I became the pitchman for All American Rose selections and quite enjoyed myself. After a couple of weeks, the strike ended and we all went back to work. Today, we believe consumers are so different from those in the past. It is not so. Then as now, gardeners knew little and were equally confused by all the new plants. If they could talk to anyone who knew anything, a sale was quickly made. They shopped for value as much as price and were no different from my daughters today.

Louie and Me

In the early Seventies, a number of events were swirling around my career and the province in which we lived. Our little home in

Pointe Claire was cozy with a lovely garden at the side. Susan was a stay-at-home mom as most mothers were at that time, and Laura was growing even taller. As my interest in gardening around the house grew, so did the interest of neighbors, colleagues, and friends. After the third or fourth request to help with their yard, plant a hedge, or advise on plants, it was evident that people actually believed I knew something. I did not. I was simply more interested than my neighbors, not more talented. I read catalogs, I could talk about the merits of at least a dozen plants, and I knew how to fertilize turf.

We enjoyed a couple of months' break from school in the summers. Money was tight—nobody gets rich on a teacher's salary, but the break gave me the opportunity to make a little extra cash. That I really did not know anything never stopped me before, so I approached a fellow teacher, Louis Santini. I asked if he would like to join me in a small garden maintenance business during the summers. On the island of Montreal, landscaping crews were either French or Italian; that was the unwritten law. Fortunately the name Armitage translated easily into French (arm-ee-taaje). Louis could not speak a word of Italian, but our names were golden. I made up some business cards, bought a little red pick-up truck, a lawn mower, and a few odd tools, and we were in business. I thought of myself then as an entrepreneur, but I was the epitome of everything wrong in the landscaping business. Today when I describe fly-by-night landscape operators, I could well be describing myself.

I learned an important lesson those summers. It is far more important to be honest and on time than anything else. We installed cedar hedges, did stone work for patios and stairways, edged beds, planted geraniums, and cut more lawns than I want to remember. We showed up when we said we would, did a good job, and were fair and honest. Referrals kept coming. I also learned another lesson: money is easy to talk about but can be difficult to collect. Chasing down money owed was the least favorite part of this experience. I still was no horticulturist, but others thought I was. Business was so good that I was starting jobs before school was finished and continuing well into the fall. This plant thing was interfering with my teaching—something had to give. Something did.

La Belle Province

The Province of Quebec was going through major upheavals in the mid-Seventies. There was always political and social tension between the French-speaking majority of Quebecers and the English-speaking minority in the province. Historically, most of the economic power in the province was controlled by English-speaking Quebecois and English-speaking multi-national companies, most notably American. In the late 1960s, a group who called themselves Le Front de Liberation du Quebec, otherwise known as the FLQ, wanted Quebec to secede from Canada. Most Quebecers had never heard of them or simply ignored them. Until bombs started to explode.

They were mostly detonated in mailboxes, usually in affluent English areas of Montreal. In the fall of 1970, the FLQ kidnapped two men, killing one of them. These circumstances resulted in the only peacetime use of the War Measures Act in Canada, providing widespread power to the military and police. These events gave rise to the Parti Quebecois— the party that believed in the separation of Quebec from Canada.

In 1974, French was declared the official language of Quebec. Perhaps this made sense in a province where, at the time, eighty-five percent of the population spoke French. However, like a long-suppressed Jack-in-the-Box, the French language became the central theme in Quebec politics. By law, services, signage, businesses, education, and the courts had to be conducted in French. Montreal was the undisputed financial center of Canada at the time, but the English power structure (financial businesses, multinational businesses) were not about to be told to conduct their meetings, or translate their minutes, into French. Many started to migrate to Toronto. If that was not enough to make businesses leave, Bill 101 was. It legislated that all store names be in French, and their English names, if displayed at all, be in letters half the size. It also decreed that anyone moving into the province from other provinces or countries, including the United States, must have their children attend French schools.

Attracting people to the city became almost impossible for businesses. So, as French power increased, the financial powers simply left. Entire office buildings were vacated overnight; people by the droves packed up and headed west. Within a few years, Toronto became the financial hub of the country.

That education was always at the mercy of politicians was worrisome for many. Job security in the English-speaking school system was now even more tenuous. The fragile nature of the system and the conflict between teaching and landscaping had Susan and I questioning whether to stay in education, or even in the province. Add to that the realization that being an excellent teacher required a major commitment day in and day out. That is true for many professions, but in teaching, no matter how good you were, there was little financial incentive. The only promotion a good teacher can realize is to not be a teacher, but an administrator. I did not see myself as a principal or vice-principal. I simply was not a desk person. So, with one child already and another in the planning stages, we looked down the road.

There are a number of forks in anyone's road of life, but some stand out more than others. And the next one was remarkable.

Six

"Behind every successful man is a brilliant woman."

. . . .

- Anonymous

The Wolves at Bear Lake

I must have been very slow, because in all this time, I never thought of horticulture as a career. Not once—not planting tulips, not cutting grass, not laying stone, and certainly not planting grave sites. It simply did not cross my mind. Every fall I was still sneezing and miserable with allergies, and I looked at the landscaping thing as only a job. However, times were changing. The more I thought about it, the more I was convinced that I should go back to school. In April of 1975, Susan was listening to me going on and on about the only course I was enthralled with in college, this thing called ecology. I had just finished reading *A Whale for the Killing*, written in 1972 by one of Canada's finest authors, Farley Mowat. Mowat was one of the first Canadian writers who brought the interactions of people, wildlife, and the environment alive in brilliantly readable prose. He made ecologists out of many of us.

I visited the same professor from whom I had taken a class. He was even more enthusiastic about his work as I asked about the possibility of graduate studies in ecology. He talked about his research studying the interaction and impact of wolves on the environment. He discussed how wolves were misunderstood and explained their importance to ecological balance. His fieldwork was near Bear Lake in northern Manitoba, not far from Hudson Bay. As we talked, I became excited with the adventure of traveling to such a remote spot to be involved in an area of science that was quickly evolving. What young man would not want to chase wolves around northern Manitoba? We chatted about going there during the summers and to other less remote places when it was too cold. I would take classes at the college in the winter. As the meeting came to a close, we shook hands and planned a time the following week to talk about details. I left the meeting pleased about the opportunity but also concerned about the time away from Susan and Laura.

"You're going where?" Susan asked. "For how long?" The conversation was not going well. She realized I was serious about going back to school and supported that idea, but didn't see me as a wolf-chaser in northern Manitoba. She had not bought into this idea at all; however, she agreed to meet the professor to better understand what I was going on about.

. . . .

We arrived a few minutes early. His door was locked. We waited around for an additional fifteen minutes, but he still had not appeared. At this point, I was restless and checked the door into the small lean-to greenhouse across the hall. Surprisingly, it was unlocked. "Look at this, Susan," I said as I pointed to the beautiful flowers of a lipstick plant. "Isn't this neat?" I exclaimed as we passed by a rex begonia. It was the first time I had been in the greenhouse with her. I was happy sharing my enjoyment of this special place with her. Suddenly, there was a moment of clarity when we both stopped and looked at each other. "What in the world am I doing chasing wolves in Bear Lake, when this is what I want to do?" She smiled, and as we left the greenhouse, I noticed the professor's door was open. We walked right by.

. . . .

I think of that day often. What if he had been on time? What if he had a plan? What if the greenhouse door had been locked? What if Susan had not come? It was as if a curtain parted and revealed a fork in the road, one I had not even looked for. Once revealed, we took it.

On our return home, we had a dilemma. What now? We were both ready to leave the province and the craziness going on there. We realized that if a break was to be made, it was necessary to rewire. The future looked like a deep abyss; walking backwards was a dead end; walking laterally would accomplish less. It was time to jump into the unknown.

When I was around ten years old, I received a small framed poem, perhaps for Christmas. It hung near my bed. It probably came from my dad or his dad. There was a good deal of Scot blood in their forebears, and it featured a small black Scottish terrier. Beneath was a poem, probably not one for the Poetry Hall of Fame, but one a ten-year-old boy memorized almost immediately. As I was thinking about that abyss, I thought of those words again:

What's the use o' howlin',
Tho' the grind is long and hard,
The path to happiness never was
A well kept boolyvard —
Jes' forget about yer troubles
That get ye riled and vexed,
Why—the spice o' life is guessin'
Jes' what's comin' next.
                                        — Cecil Aldin

Ontario

My fascination with wolf research had whetted my appetite for additional schooling—this time in horticulture. However, a few tiny problems surfaced. We had little money, and we were expecting our second child in July. In Canada in the mid-1970s, only three schools offered advanced degrees in horticulture. One was MacDonald College, but as it was in Quebec, we crossed that off. The others were the University of British Columbia and the University of Guelph in Ontario. Hasie and Phyllis were living in Waterloo, Ontario at the time, a mere hour away from Guelph. We decided to see them and then visit the university.

I had telephoned the horticulture department and by random good luck, spoke with a young professor who invited us to visit. In May, we jumped in the car, and seven hours later were in the lovely town of Guelph in southwestern Ontario. I warned Susan that nobody is invited to do postgraduate work without paperwork, interviews, and writing tests. "We will just look around and see if this is for us."

What I meant was, "What the hell am I doing here, and where will we go if this guy is not interested?" A long trip back after rejection was not something I looked forward to. We arrived at the horticulture building, and there I met Dr. Jim Tsujita, one of the major influences in my journey towards horticulture.

I exited the car, and we started chatting. After a few minutes, he abruptly halted and asked, "Where is your wife?" I gestured to the car, and he opened the door, shook her hand, and told her that he would like to show us around. We went through greenhouses and labs, and his enthusiasm for his work was obvious. "What do you know about roses?" he asked. I told him about my times with 'Mr. Lincoln'. He said, "I just received a grant to study the influence of lighting and nutrition of greenhouse cut roses—are you interested?" In a matter of a few hours, we went from abject uncertainty to being part of an academic community. Not only that, I was offered an assistantship that would help pay for groceries. I was nothing more than a lowly graduate student, and a rather old one at that, with a three-year-old daughter and another child on the way in July, but I felt we had hit the jackpot.

The hardest part for Susan was leaving her family and our little house, which she had transformed into a gingerbread cottage. We had to sell the house, find another place to live four hundred miles away, and start all over again. Which is what we did.

The hardest part for me was leaving my students. We had a few farewell gatherings, but I knew they would get on just fine without me. However, I still have the broken canoe paddle signed by Outers Club kids, I still have some of the thoughtful comments they left in the yearbook, and, believe it or not, I am communicating with a number of them today through Facebook. The other difficult part was leaving my mother behind.

. . . .

My mother defined the word tough. Not tough like the leather on a shoe, but tough in character. In 1968, she was diagnosed with mouth cancer—hardly surprising when she smoked at least a pack of Du Maurier cigarettes every day. She went through some aggressive radiation treatment and, after a year, she was "cured." She had a minor speech problem and had lost a lot of weight, but was able to resume her normal habits. Unfortunately, one of the normal habits she resumed were the Du Mauriers—she simply couldn't kick the addiction. A few years later, the cancer returned with a vengeance, and she had to have radical surgery, removing half her jaw and other assorted pieces. Such a beautiful woman—now looking a bit like a ghoul and having to learn to speak all over again. When we visited with the kids, they took a little time to warm up to this rail thin, odd-looking woman. However, soon we went for walks outside and had tea and conversation inside. Her apartment was just like the house had been—a ton of books falling off tables, talcum powder all over the bathroom, and a drink in hand. This time the drink was Ensure Shakes®, with a little dollop (perhaps not so little) of Canadian Club® whiskey. The Ensure was her nutrition; the whiskey just helped it go down. Regardless of her pain and her difficulties communicating, she always was in good spirits. Nothing could make her complain; she never belly ached about anything and was always concerned about the plights of others. She was a beacon of tolerance in a world of complainers. Eventually, the cancer got her, and she passed away in October 1988 at the age of seventy-three. The kids still recall Grandma "Tige" with great affection.

. . . .

Mom had great patience with her sons and friends, and she passed that on to me – in most things. However I have no time for pessimists or complainers, for those who won't do something for themselves, or for people who wail in anguish and wait for someone else to take care of them. Blame my mother's stoicism for my impatience.

Life Gets Better

On July 20, 1975, Heather was born. From day one, she was her own little person. Heather was put on this earth to keep everyone straight on life's activities. If we wished to know anything about anything or anybody, we simply asked Heather. As the middle child, she constantly reminds us that the filling is the best part of the sandwich. By the time she was five, she had spoken more words than the rest of the family combined. We knew that Heather would assert herself early when she was sent to the hospital immediately after being born. It was there her reputation began.

. . . .

Heather was born active. She entered the world with lungs full of air, which she emptied in deafening fashion at every possible opportunity. When she was delivered, she was a healthy child but soon developed a rather high fever. Susan was beside herself because of Shannon. Due to her anxiety and the physician's concerns, Heather was transferred to Sick Children's Hospital in Montreal. Watching your two-day-old child taken to a strange hospital, where only the sickest of children were treated, is very traumatic for any parent; we were no exceptions.

When I visited, one of the things that struck me as strange was the lack of child-type noise in the hospital. Machines were humming, and an occasional monitor was buzzing, but few children were crying. Perhaps many were fighting so hard for their lives that crying simply was not important enough to waste time or energy on. There was, however, a notable exception. Over in the corner, off the beaten path, was Heather—telling the world she did not belong there. Her obvious tenacious hold on life was a joy to nurses, who were used to children barely clinging to it.

One of the reasons I went to the hospital was to deliver her food. Heather was only two days old, but Susan had immediately started to nurse her. Since mother and daughter could not be united, it was up to Dad to play milkman and deliver the breast milk. I had delivered many things in my day, from newspapers to chocolate bars, but I never had any inclination to deliver breast milk. Ah, the immaturity of youth. I was so embarrassed the first time, taking bottles of newly gathered breast milk in a Styrofoam container through the streets of Montreal. I was sure I was the only father who had ever done this and equally sure everyone knew the exact contents of the "picnic basket." I delivered the white gold to one of the nurses, who simply said thank you, took the container, and left me standing there feeling very embarrassed about my embarrassment. I continued my deliveries, but within a few days it was obvious Heather belonged with her mother. Ironically, not only was Heather the healthiest of all the newborns, she steadfastly refused to breastfeed anyway.

. . . .

They Multiplied Like Zucchini

In fall of 1975, Susan, Laura, new baby Heather, and I went west and found ourselves on yet another fork in the path. The college experience is quite different when experienced as an immature twenty-year-old kid or as a married man with a family. I became an A student overnight. Not that I was any smarter; in fact, I was taking advanced classes in subjects such as calculus, plant physiology, and anatomy I had never taken as an undergrad. A little maturity goes a long way. I still had no idea what I wanted to do, but I knew that I wanted to be in the plant business. I believed that by meeting professors and students in the field, I would learn so much. More importantly, I wanted to make contacts and find a good job somewhere in the country. A Master's Degree was simply a means to an end. I wanted to suck it dry for the two years I would be at Guelph.

Student Ron Dutton, left, and Dr. Jim Tsujita at University of Guelph

I had never done serious research. It rather frightened me. Scientific method, experimental design, data collection, statistical analysis, and journal writing all intimidated me. These tasks, however, were made more pleasant by being in a warm greenhouse on a wintry day. Eight-feet-tall red roses, some growing even more strongly by the addition of special grow lights, dwarfed us. Soon I was cutting roses for Thanksgiving, Valentine's Day, and Mother's Day. Every rose was measured for stem length, and the yield of every bush was calculated. We had roses in the bathroom, in the hallway, in the girls' rooms—the bloody things multiplied like zucchini, and soon we resorted to putting them in the mailbox for the postman and watching the garbage men clutch bouquets as they hauled off the trash.

Page after page of data was collected. I started writing about the research. Scientific writing has its own rules and sentence structures and is unlike any other literary genre. However, as I concentrated, I found I enjoyed the discipline of writing. This turned out to be a very good thing indeed, because the academic world is all about writing. Preparing grant proposals and publishing in refereed journals have always been the law of the land. Regardless of teaching skills or the number of magazine articles written, if papers are not forthcoming, you are gone. It was that way then, and even more so today. As a lowly graduate student, I did not feel that pressure, but I was expected to write a thesis in order to graduate. My enjoyment of writing was a bonus.

Early on at Guelph, I became quite interested in foliage plants. The green movement in malls, office buildings, and homes was gaining momentum. I learned all I could. I travelled to the Foliage Show in Orlando (now the TPIE) to learn as much as I could. Dr. Tsujita recognized a crazy man when he saw one and encouraged me to learn about as many subjects as possible. This was unusual advice from a major professor—most wanted their students to concentrate solely on their research so more papers could be written. One of the professors at school was Dr. Doug Ormrod, an excellent teacher and well-respected researcher. He was interested in the use of controlled environmental chambers for research. He had written a one-hundred-page manual concerning operation, maintenance, and usefulness for environmental research. I was so impressed that he had written such a tome I asked how he did it. He replied, "I write one hour a day, every day." That sure sounded simple, but I knew if it were that easy, everyone would be a writer. I took the advice to heart and think of it whenever I am writing an article, a paper, or a book. When students ask me how to become a writer, I simply repeat Dr. Ormrod's counsel.

This came in handy when Dr. Tsujita asked if I would like to write a manual for the Canadian government about the use of foliage plants for office buildings. He did not have the time or desire and passed it on to me with the promise of $500.00 payment from Ottawa. I researched light levels, nutritional needs, soils, container sizes, insect and disease control, and sources—information that would be useful if plants were to be used in government buildings. I sent it in, someone said thanks, and that was the end of that. However, it was my first "book," weighing in at about 180 pages, and, more importantly, my first paycheck for writing.

We all settled into a nice little condo and were very much enjoying Guelph. I even became the "Plant Doctor" and advertised in the local paper that I would make houseplants like new again. Garbed in a lab coat and armed with a shaker of Osmocote® and some spritzers of plant shine, I made house calls in the evenings and weekends. I was offered lots of tea, a beer or two, and found that people who cared enough to have someone look at houseplants were usually nice people. I made a few dollars while learning that people really cared for plants— something I have found to be true time and time again.

As our time at the university was coming to an end, reality set in. I had to find a job. With two small children and a mountain of debt, it was time to earn some money. I had planted a small garden and had become more aware of the potential for urban gardening. There were more people buying, planting, and decorating yards and patios. More trees were being placed in town squares and parks. Unfortunately, the term "urban horticulture" had not yet been coined. It was obvious to me that there was a great opportunity for research and service for the department. I approached the department chair and suggested that this would be a growth position for the university and department. I may have mentioned that I would like to apply for it.

He squinted at me and muttered, "Urban horticulture?" Horticulture was still very much tied to fruits and vegetables; the real potential of the ornamental side was just being realized in progressive universities in the States, but not so much in Canada. Unfortunately, he was not interested. I suppose my idea was exonerated years later when nearly every horticulture department instituted some form of urban horticulture. However, at that time all I knew was that my job prospects had diminished considerably. Then some good news arrived.

Pierre and Maggie

I was asked by the Canadian government to apply for a job in Ottawa. My report on foliage plants must have been exhumed and actually read. Somebody decided that it was time to bring green living plants into the government buildings, and they needed a person who knew a little about plant selection and maintenance. Not only that, the job also involved working in the greenhouses in Ottawa and growing flowers for state and international functions for the Trudeau government. Pierre Trudeau and his wife Maggie were Canada's rock stars; the entire country loved them. And better yet, Maggie loved flowers. She believed they should be grown on site, not brought in from overseas. It seemed the job description was written for me. I would have people working for me, following my instructions while being paid a handsome salary. I visited Ottawa and was walked around building after sterile building, each yearning for something green. I hiked through the greenhouses realizing that this was a huge job—but one I was ready for.

I was excited, but Susan seemed to have some concerns. I was not sure why, but she said there was something that just did not feel right. Perhaps she had reservations about working for the government or going to Ottawa. She asked me to put off signing a contract until we could go to the city together. We drove to Ottawa—a beautiful city, clean, efficient, and a wonderful place to raise children. It was only two hours from Montreal and family. Susan and I were sitting on a bench, looking downtown. She had been rather pensive, then she looked at me and said:

"Where is your office?" I pointed to a tall building down the road.
"What floor will you work on?"
"The nineteenth," I replied.
"Will you have to wear a tie?"
"Yes," I said
She looked at me and said, "Don't take the job."

I remember this like yesterday because it was so alien to what I thought we were trying to do; that is, go back to school, get a job, and raise the family. She calmly pointed out that I should not work in an office, as a manager, surrounded by other workers in ties. Essentially she said (more politely) that I would make a lousy administrator and would hate the job in a year or two. She tried to tell me that my strengths were with people, with teaching, and getting my hands dirty. All this time we were sitting on the bench, she was explaining to me who I was. It was kind of embarrassing. But she was right. She said, "Something else will come our way—this is not the right choice. You will be miserable." And just like that, I turned down the job.

It was June of 1977. I had earned my master's degree in horticulture, turned down a lucrative career opportunity, and my job prospects were not exactly robust. Much like the light bulb in that little greenhouse at MacDonald College, I was so focused on getting a job and making money that I didn't recognize something right in front of my face. I was really enjoying the university atmosphere. I thought about how much I enjoyed learning about plants and physiology, about the people in ornamental horticulture industry, about ... and eureka! I realized that this was where I wanted to stay. In the beginning, I was absolutely terrified by doctoral students and professors in their labs. I did not believe I was capable of doing research work, and certainly not competent enough to write a paper reviewed by scientists. When I arrived at Guelph, I felt out of place and intimidated. Two years later, I was still in awe of professors who had "made it," but I was no longer intimidated. I wanted to be a professor in horticulture.

Seven

"You can't make decisions based on fear and the possibility of what might happen."

. . . .
- Michelle Obama

So Now What?

There was no future where we were, no prospects in sight, and it was too late to apply to a university elsewhere. People entering a PhD program in September had applied six to twelve months ago. At this point, the chances of being accepted anywhere were slim to none. But reality never stopped us before.

While I was writing my thesis at Guelph, I reviewed existing literature about lighting and nutrition; that is, I went back through journal articles to determine what had been published on the subject. I cited a number of papers on light and roses, some published by Dr. W. J. Carpenter and Dr. W. H. Carlson at Michigan State University. Dr. Carlson was also working on bedding plants, a quickly emerging group of ornamentals. I had done some research with a new plant, the seed geranium, while I was at Guelph. In fact, my very first refereed publication concerned the influence of supplemental lighting on their growth, and I noticed this fellow Carlson was working with seed geraniums as well. There was nothing to lose, so I decided to call him.

I explained the situation to Dr. Carlson and was pleased he did not immediately hang up. He was most understanding and said he would check with his colleagues to see if anything was available. He was also quite clear: "Do not get your hopes up, it is very late."

I must have a silver cloud hovering over me, taking care of me in times like this. Dr. Carlson called back the next day. He asked if I could come to East Lansing for a visit. I left Susan and the girls behind. I drove through London, on to Windsor, across the bridge at Detroit, and finally arrived two hours later in East Lansing. If I did not blow this, not only would I enter a prestigious university, but I would also be in the same town as Hasie and Phyllis. Since Phyllis had straightened him out, my brother had earned his MBA at the University of Alberta. He was now doing his PhD in Accounting at Michigan State. This would be the first time we would be living near each other since we were at MacDonald College.

When I reached the campus, my eyes bugged out. The student population at MacDonald may have been two thousand on a good day, and few cars were seen on campus. The town of Sainte Anne de Bellevue had no more people than that, and there may have been a stoplight in town. Everything was so close we walked everywhere, often times from campus to our apartment. By comparison, the University of Guelph was huge, boasting nearly ten thousand students and an equally busy town. Cars were needed to come to work, but were only permitted around the campus perimeter. On campus, we walked to class and nearly every-where else. Michigan State University was a megalopolis.

The first thing I noticed when I drove onto campus was a stoplight! A stoplight for goodness sake; what kind of campus has a stoplight? The place was enormous; I drove around for nearly thirty minutes, totally mystified, and finally found a parking lot near some flowers. They turned out to be part of the Trial Gardens at the Horticulture Department. I met Dr. Carlson, and we chatted about my grades at Guelph, my research, my writing endeavors, and my interests. He introduced me to graduate students and professors and showed me around labs and greenhouses. We walked through the Trial Gardens, of which he was immensely proud. I was in a trance; I had never seen such facilities, so many labs, and so many students. I bet there were at least thirty graduate students in the department; he alone had three.

Dr. Carlson was excellent at many aspects of horticulture, but he was passionate about the promotion of bedding plants. He spread the message across the Midwest and the country. He was as close to a hellfire and brimstone preacher as there was in the industry. In the late 1970s, the extension service was strong in many states—Will was Mr. Extension to the greenhouse industry in Michigan. He would visit growers when problems arose and give them hell if they were not following proper grower protocol. Some of the best growers in the country are in the state of Michigan, and much of their increase in the market share and recognized quality could be credited to Will Carlson.

So it was no great surprise when Max Koppes—a strawberry grower from Watsonville, California—approached Will. Carlson had been offered a grant from Koppes, who wanted people to know about the ornamental value of strawberries. He felt he had a particularly out-standing product and wished to increase his reach into the greenhouse/gardening market. Will had not taken up Max's offer because he did not have time or anyone to do the project. My phone call came at a most opportune time; he called Max, negotiated the money to do the work, and suddenly I was about to become a strawberry researcher. Based on the grant from Koppes Farms, Dr. Carlson offered me an internship. He also made it clear that the strawberry work was to be independent of my PhD research. In other words, I would have two major research projects. I told Susan, "I thought about it for about ten seconds, and then said, 'Yes.'" She claimed her tears were tears of joy, but I am not so sure there were not a few "here we go again" tears as well.

Dr. William Carlson,
Michigan State University

Persona Non Grata in the U.S. of A.

Dr. Carlson assured me that the application process would be routine; my grades at Guelph were good, and I had excellent references. Unfortunately, I forgot to tell him about my grades at MacDonald. I also neglected to fill him in on the Canadian grading system in the 1960s, where my seventy-three percent was pretty darn good. The MSU registrar did not think so. Dr. Carlson had to personally guarantee my good performance to get me in—as if I did not have enough pressure, that's all I needed. There were also immigration issues. As a foreigner, I had to apply for a student visa for my family and myself, so I could go to school and, more importantly, accept the assistantship money. We were trying to get rid of the condominium, tie up loose ends in Guelph, find a place to live in East Lansing, and raise two precocious daughters. The department helped a good deal, but everything took much longer than I thought it would. Perhaps I should have been more patient.

First one week passed, then another, and another—still no visa. It was September, and with classes starting the next week, I was anxious to get there. I wanted to find a desk, meet other students, and otherwise get my feet on the ground. So I decided to sneak into America.

I did not see a problem with the border. I would take a bus from Guelph to East Lansing and tell the people at the border that I was visiting my brother, which was true. I wouldn't mention anything about school. Anyway, I would return home that weekend and pick up the visa along with my family. Susan, who is an assiduous rule-follower, was against the idea from the start. "You need a visa; something will go wrong."

The last bus left about four in the afternoon, and the ride was uneventful until we reached the Detroit border crossing around seven that night. The bus parked, then a border guard asked that identifications be sent to the front. This was before passports were required between the two countries, so I passed along my drivers license. About five minutes later, a couple of serious-looking fellows boarded the bus and started asking everyone questions. When my turn came, I started with my rehearsed story. He sternly asked, "What is your brother doing there?" "He is a student at MSU," I replied.

"What is he studying?" and many more questions followed. Then he looked me in the eye and asked, "Are you planning on being a student there?" All eyes of my fellow travelers were on me.

I am a lousy liar. Lying to this officious fellow did not seem a particularly good idea. So I mumbled something about being accepted to MSU, trying to get a head start on classes. "What did you say?" and with that hauled me off the bus and took me for interrogation into the customs building. They grilled me about this and that; they even made an official file on me. The people on the bus were stuck at the border. They had to wait to see if I would be returning. I would not.

The guards were most emphatic, belligerently so I thought, that I needed a visa to get to my destination. My arguments about being a good Canadian neighbor fell flat. My insistence that the visa was in the mail only convinced them even more that I was intent on doing harm to the country. When, after another half hour of intense questioning, one of the guards gestured outside and the bus pulled away, I knew I was in trouble. When he closed the file and asked, "Who can you call to pick you up?" a bad situation turned worse.

I called my good wife and said, "Susan, I have been thrown out of the country. Can you come and get me?" Silence.

"You what?" Four hours later, Susan pulled up to the border building and I was handed over like a criminal to his warden. All that was missing were the cuffs.

Don't Hang Any More Curtains

We drove back to Guelph, she never once said, "I told you so." I think she was too mad. About half way home, she looked over at me and said, "I called Hasie, and first he said that your trying to sneak into the USA was the dumbest thing he ever heard of," and then a smile crossed her face as she repeated my sage brother's reaction. "He burst out laughing when I told him what you did." And we both started to laugh out loud. "When the kids get older, they are going to love this story."

Thus concluded my first extended visit to the United States of America. Visas and other official papers eventually arrived, and the family Armitage was permitted to live in Michigan. I joined the graduate student contingent and found a desk in the graduate room, where there were not thirty, but seventy-one other students—all in horticulture! That was bigger than my entire agriculture class at MacDonald College.

I realized I had landed in roses once again. Everyone made me feel at home, showed me around the campus, pointed out lab locations, and shared many classes. The graduate room became a study hall in the evening, and every night would find grad students working on their research or classes well after midnight. We spent three years in Michigan, and we recall most of them fondly. However, there were times when I realized my pursuit of higher learning might not come to be.

"Susan, don't hang any more curtains!" I had returned home from my first quiz in my first class at MSU. I knew I was in over my head on the first day the professor started talked about spectrophotometers, magnetic resonance, molecular bonding, pectins, amylases, and enzyme theory. I looked around the class and I realized I was not only clueless, but by far the most clueless of anyone there. I had no business being in a high-level plant physiology/biochemistry class, but there I was. The professor's name was Dr. Zeevaart. He showed no mercy. I received the lowest grade in the class on the first midterm. I was mortified, but mostly I was scared. I had done the best I could, and my best earned me an F grade. I was in deep trouble.

I studied. I read. I begged my smarter colleagues for help. I put off everything other than classes that quarter. Lord be praised, I eked out a C. I felt as excited as I did when I received a fifty-five percent on my English class at Mac. Other classes were equally challenging, but I was better prepared. I realized that the more I learned, the more I knew I could not learn it all. So many opportunities were available that classes became a background noise, things you were simply expected to succeed in, but not the reason you were there. The reason I was there was to gain as many experiences as possible in the industry, and, most importantly, conduct research that would allow me to graduate and take care of my family in the future.

Graduate students are like rats in a maze. The maze keeps changing. Unless you lift your head every now and then, you will forget why you are in there. A master's degree is a filter designed to provide a path for the future. Once completed, you find a job, or continue to the next step in academia, much like a preseason match in an athletic event. Most master's students arrive with little experience in unique research and creativity. The major professor often assigns a specific research task and provides ample assistance in research protocols, writing, and experimental design. At least that was the case with Dr. Tsujita at Guelph. Not so at MSU.

You're On Your Own

It turned out that Will Carlson was exceptional at organization. He was genius when it came to finding money. He was also excellent in bringing additional opportunities to his students. However, he was hands off on the research side. Essentially, he said, "Bring me your proposal, and we will take it to your committee." The topic, the protocol, and the writing were up to me. That was a rather daunting task for someone who was still trying to find the greenhouses. It became immediately obvious that if I were to succeed, I needed to take the initiative—one of those initiatives being to find people who could help and guide me.

The graduate students and I working under Dr. Carlson became very close, as we all found ourselves in similar situations. I found approachable professors to serve on my committee and help out when needed.

It was well known that Will was a believer in the "sink or swim" philosophy of success, but I didn't know any of this then. Later, when I was applying for a job, one of those interviewing me was an MSU graduate. He said. "Allan, with Will Carlson as your major professor, you will be either very good or very bad."

I have nothing but respect for what Will brought to the university and to the industry. He was a giant in the industry in Michigan and nationally. By sheer hard work, he brought new innovations, the latest in greenhouse structures, and ideas for research that made MSU the undisputed leader in horticulture research and service. He was an idea man, not a hand-holder. I was on my own.

Fortunately for me, MSU attracted the cream of the crop in graduate students. The faculty, the program, and the reputation of the university drew young men and women from all over the world. Once we were there, we were all in the same boat, trying our best to get through classes, understand the concepts of statistical analysis and experimental design, and work on unique research projects. All of us knew the stakes. We circled the wagons to help each other out when possible. People certainly helped me. Colleagues such as Carl Sams (University of Tennessee), Miguel Leon (Madrid), Gene Mero (vegetable breeder, CA), Bill Randle (University of North Carolina), Bill Wolpert (UC Davis), Denny Werner (North Carolina State University), and many more came to my aid. Although we did not realize it at the time, this was probably one of the high point years for graduate student education at many horticultural programs in the country. By the way, we had strawberry baskets hanging on our windowsills, strawberry containers by the door, and one happy postman when we hung strawberry-laden baskets from the mailbox. We also made Max Koppes happy when we published and spoke about the potential of strawberries for ornamentals. This work put food on the table.

Why Aren't They Flowering?

We were doing just fine at MSU. Classes and research were going well, and we had found a place to live close enough that I could ride my bike to work. The main roads in East Lansing were always busy, with cars entering and exiting the main roads on their way to work or home. My route took me along Grand River Ave, one of the main thoroughfares that bordered the campus. Riding a bike in town was not the safest means of travel, particularly at 8:00 in the morning. In 1980, no one even thought about bicycle helmets, let alone wore them. Biking to work was inexpensive, and I had been riding without incident for many weeks. Not this day.

I crossed a street adjoining Grand River. Before I knew it, I was flying through the air, bike mangled on the road, and me coming down to earth. I looked pretty rough, and I felt much worse, but I was conscious and could talk to the EMTs before being rushed to the emergency room. After groggily answering who I was, what I did, and other questions, I was put into a wheelchair. I was in no danger, but I was in pain. Fortunately, my head was not badly hurt, but the rest of me ached all over. In such a state, I began my journey to X-ray. We seemed to take a lot of turns, but I was half comatose, so nothing should have surprised me—but this did

. . . .

After what seemed an eternity, the nurse stopped the wheelchair in front of a fluorescent-lit table of African violets. I slowly raised my head to look at this blurry scene, and she said, "I see you are a horticulturist. Can you tell me why my African violets aren't flowering?" (I can't make this up.) In my haze, it dawned on me that she was serious. She even handed me one in a pot—sure enough, it wasn't flowering. I mumbled something about light, fertilizer, and dividing crowns; it must have sounded acceptable. She thanked me, put the pot back on the shelf, and wheeled me on my way.

. . . .

Other than aches and bruises, the only injury was a broken right wrist. It was placed in a cast, and I was home that afternoon. I looked awful—purple, black, and blue, and my wrist hurt like crazy. All I wanted to do was sit in a chair and not talk to anyone. My daughter Heather, however, had different ideas. She told her friends at school about her dad's flight through the air, then charged them ten cents to come in and see me.

I was very lucky that it was not much worse. Susan and I smile when we think about it now. Not only that, but I met a passionate gardener, and my five-year-old daughter earned a few dollars. I also found walking to school was quite pleasant.

Captain Hook

Heather was growing quickly, and Laura had started kindergarten. As expected, Laura was tall for her age. When she turned four, she was often seen around the house looking like a figure from Peter Pan.

. . . .

Laura sucked on a pacifier when she was a baby. To Susan and me, that pacifier was the most magical piece of rubber ever invented. It was the only thing in the entire world that could quell her crying, particularly when she would wake at night. On the other hand, I can't count the number of nights I crawled half asleep under her crib to find that stupid thing and plant it back in her sobbing mouth. As she became older, she replaced the pacifier with an empty plastic baby bottle.

At first, this seemed good because it stayed in the confines of her bed and I wasn't groping under the crib quite as often. It was simply a big pacifier. I believed she would give it up after a reasonable amount of time, but was I wrong. Not only was the bottle her pacifier at night, she would also walk around with it during the day. Most of the time the thing was empty. Made of plastic, it was scratched, discolored, and altogether quite disgusting—the kind of thing only a child could love.

When Laura was three, we consulted Dr. Spock. Like all parents of first children, we actually believed him. The good doctor told of the great damage a bottle wrought on teeth and palate. Doubtless, our child would grow up with no teeth, distorted lips, and other permanent damage, all because of a fifty-cent piece of plastic. With tough love, we decided to take it away! After enduring three solid days and nights of crying and carrying on, Plan B seemed to be in order. Using my superior intellect, I told her we should cut the bottle in half so she could use the bottom part for a cup and throw the top half away. Laura thought this a grand idea.

We went down to the dingy Pointe Claire basement, and, with the old hacksaw, we began the surgical procedure. Her eyes started to bulge a little as we cut through but when finished, she gingerly placed the top half in the garbage pail and took the "cup" upstairs for some juice. Mom and I patted ourselves on the back. This child raising stuff wasn't so tough after all.

The next day, Laura emerged with one arm longer than the other. She had fished through the garbage, retrieved the top of her bottle and jammed it over her right hand. She looked just like Captain Hook—except he didn't suck on his hand. We were hardly pleased. Watching a three-and-a-half-year-old, three-and-a-half-feet-tall child running around the house sucking on an artificial hand hardly built confidence in our parenting. People looked at her a little strangely, and we were afraid that when she started kindergarten, she would be thrown out for wearing a bottle (a definite breach of the dress code). It went everywhere she did, from trips in the car to trips to bed. It was near and very dear to her.

In Guelph, she announced that on her fifth birthday, she would throw the bottle away. This was a big decision. Although we wanted to saw the thing off her arm, we agreed to wait. Every other week we reminded her of her promise, and she would nod solemnly and then go about her business. The morning of her fifth birthday, she quietly walked to the garbage pail and gently placed the bottle into the bin and out of her life. Watching her at that moment, Susan and I realized that everything has its time. No amount of "parenting" is going to change it. We were so proud of her; our eyes welled up like we were in some sappy movie. From then on, we knew Laura had vast inner strength that would serve her well later in her life.

. . . .

The Horticulture Revolution

The assistantship stipend was modest but most important to our well-being. Dr. Carlson asked if I would take over the day-to-day running of the research greenhouses that were used by his students. The greenhouse complex consisted of about three bays, some bays further divided into different temperature regimes (i.e., a warm house and cool house) and a number of environmentally controlled chambers. Research on temperature and light manipulation could be carried out in both the chambers and the greenhouses. This was terrific for me since my research and that of my colleagues took place in those houses. Being there allowed me to keep a close eye on everything. It also provided a bit more money. I would take my Walt Disney lunch box to work, care for the geraniums, marigolds, petunias, and kalanchoes students were growing, then go to class. I had been in school for most of Laura's life, so when she started kindergarten in East Lansing, she was quite happy to be "doing what daddy does."

It appeared that I was on my way to discovering the great world of plants. As a graduate student, the path was well defined—study, write, and publish. Although struggling financially, we were making enough money that no one was starving. Dr. Carlson was involved in many industry innovations, one of which was the commercial growing organization called Bedding Plants Incorporated (BPI). Started in 1968, it became one of the most influential organizations in the greenhouse industry. I became involved, thus meeting many researchers, growers, and extension people. In the 1970s and into the 1980s, the extension service in many states was at the forefront of education and change in the industry.

Today, extension is but a shadow of itself, too often replaced by bulletins and private consultants; however, at the time, thanks to Dr. Carlson, my education expanded exponentially from the lab and research bench to the real world of production and scheduling.

The industry was growing so rapidly; major developments in production efficiency and plant scheduling were high on the list of research topics. Computer controls and marketing were not even on the radar. Topics like plug technology, soilless mixes, High Intensity Discharge (HID) lighting, and growth regulators were just emerging. Breeders were introducing new cultivars of bedding plants, but the concept of "new crops" was years away. Perennials had been around for years and were just becoming "mainstream." Such were the heady days in which I found myself. I was immersed in a horticulture revolution.

The Last Hurdles

Heady days aside, I was just one of many graduate students. I had research to accomplish, a dissertation to contemplate, courses to pass, and the terrifying oral examination to survive. The oral exam is an example of "what does not kill you makes you stronger." We sit in a room in which professors line up like a firing squad and shoot questions at you about any and all imaginable subjects. To be fair, they try to only injure you, not kill you.

. . . .

If one truly wants to experience pain and agony, try sitting through a three-to-four hour pasting from a bushwhacking gang of professors, some whose sole reason for attending is to shoot off their guns of knowledge for all others in the room to hear. Each takes a turn trying to impress the others, while at the same time seeing how much the student can squirm. The squirm factor is very important in an oral examination. The student needs to show a respectable amount of squirming in front of his elders, but not too much, or the jackals start baying.

. . . .

Having survived this particularly macabre form of higher education, I was delighted when a friend offered us his lake front cottage for a few days. When we all set off to the cottage, our minds were on sun and surf, not on our accident-waiting-to-happen daughter. Big mistake.

. . . .

We were relaxing while the children enjoyed running in and out of the placid waves. All seemed well until Heather started screaming. Initially, we thought she had simply jammed her toe. After all, this was Heather. On lifting her out of the water and seeing the blood flowing from her foot, we realized this was more than a stubbed pinky. With her foot wrapped in bloody towels, we raced off to the county hospital, an hour away over hilly, twisting country roads. The drive was absolute hell, as night settled around us. We didn't know where we were, where we were going, or how to get there. The towels were red, and Heather was moaning the whole way. It turned out she had stepped on a broken clamshell and suffered a particularly deep and ragged cut. When we finally arrived at the small hospital, she was whisked into the emergency room with Susan at her side. Laura and I sat outside the room— I will never forget Heather's pain that night. The foot is a particularly tender part of the anatomy, and when the doctor probed and cleaned it, she was beside herself. Stomach-wrenching cries accompanied each injection of painkiller around the cut.

Laura and I were convinced that her foot was being amputated. It was all I could do not to race into the treatment room and rescue my daughter. The yells and screams soon subsided to a background of sobbing, even harder to listen to than the screaming. She finally hobbled out on one crutch, subdued but not defeated. The appear-

ance of a tiny smile lit up the hospital corridor like quiet fireworks on a still Michigan night. As we headed back to East Lansing, our holiday was shattered, but our daughter was intact.

. . . .

Canada Said, "No"

My research in the environmental influences on marigolds and geraniums, as well as work with anthocyanins, auxins, and gibberellic acids, provided grist for publications. I attended, and spoke if invited, to scientific meetings and industry shows. I had learned a few things from Dr. Carlson about the academic world. One was the age-old adage "publish or perish," and the other was "get out of your seat and meet people." I tried to do both.

Before I knew it, three years had flown by. Once again, we were at a crossroads—where would the next path take us? We very much wanted to go back home to Canada. When we went to the States, the last thing on anyone's mind was staying. We loved Michigan, but Canada was home. However, when I looked for positions in Canadian universities, the cupboard was bare.

In 1980, Guelph was not hiring, the University of British Columbia had nothing, and there were no jobs in Atlantic Canada or the Prairies. The Canuck door was shut. We were both very frustrated, but soon another event occurred to provide even more urgency.

One fine morning in March, Susan raced to the bathroom. After a few days of this and a trip to the doctor, she announced that wherever we went, we would soon have another baby to feed. I was to be job-less once again, but being broke and scholarly was no longer an option. I probably could have found a post-doc in the department, and certainly I could have worked with Dr. Carlson for a few more months, but neither of those was particularly appealing. I was ready to find my own way; the problem was there were few ways to be found.

Luck was on our side again. If we were to stay in the U.S., one of the few places I had considered was Ithaca, New York. In May, Cornell University asked me to apply to a position coming open in ornamental horticulture. Here was a great university, an excellent department in a beautiful part of the country, and not far from home. We were both excited and ready to pack our bags. The application was submitted.

We were ecstatic a few weeks later when the department indicated great interest, but a funding issue had arisen, and the job would not be opened for at least another six months. They gave me assurances that I was the number one candidate and that the funding issue was temporary. We said we could wait until the fall.

Meanwhile, another institution also suggested I apply. I confess I had never thought about, in fact I knew nothing about, the University of Georgia. While we awaited further word from Cornell, the department chairman at UGA, J. B. Jones, urged me to apply. What did I know about Georgia? Good grief, Tara, cotton, grits, and heat. And what kind of a name was J. B. anyway? He told me that in a few years, the UGA horticulture department would be one of the finest in the nation. They were filling positions with the best people they could find and asked me to come down and take a look.

This was impressive stuff. Even though my countrymen had scorned me, two well-established universities in the United States were interested in my joining them. It is a well-known fact that the more education one acquires, the fewer the jobs available. Positions at universities are difficult to find at any time, and this time was no different.

I called Cornell again, and nothing had changed or improved. With Susan sick every morning and two rambunctious daughters at home, it was silly to not at least apply. When some of the faculty at Michigan State heard about this, they let me know in no uncertain terms that UGA was an excellent university and that indeed the department was making a determined effort to be one of the best. A week later, I was asked to interview.

Eight

"Buy her a diamond, get a free hunting rifle!"

. . . .

- Roadside ad near Atlanta, Georgia

Next Stop: Athens, Georgia

I first set my feet down in Athens, Georgia, in early summer of 1980. The town was small but rather charming, and the university was bustling with about 25,000 students. The university was beautiful, with enormous oaks and magnolias and history that oozed from every pore. Buildings everywhere on South Campus, where the horticulture department stood, were being renovated. Additional structures were going up, and the area was bursting at the seams. I was interviewed, toured, and introduced to staff and deans over a couple of days. I was fascinated. I met Dr. Stan Kays, a recent graduate from MSU already making his name in postharvest physiology; Dr. Max Vines, a plant physiologist who loved horticulture; Dr. Harry Mills, making breakthroughs in plant nutrition; and a fellow they had recently stolen from the University of Illinois, Dr. Michael Dirr. Everyone was positive and excited about the future. It rubbed off on me.

Upon returning to Michigan, I told Susan about the people, the campus, and the town. I laughed about the ubiquitous Bulldogs—in stone, on billboards, and on campus—I even saw grandmothers barking. I must have done something right, because I received a job offer not long after. No one was knocking my door down from Canada, Cornell was still a misty maybe, and we really could not wait. Faculty and fellow students were pleased and chuckled about a Canuck possibly going to Georgia. "The heat should not be a problem for you. Once it goes over eighty-five, it is off your scale." I should have been so lucky.

When we left Canada, we had no intention of working in the United States. And certainly the South was the last place we ever thought we would find ourselves. We knew as much about Georgia as people in Georgia knew about Canada; that is to say, embarrassingly little. As in most changes in a family's life, the children assimilated the news much better than the adults. Laura was eight, Heather was five, and Jon was just a bulge. Susan and I talked. We decided to treat the move as two-year foreign assignment. Working in Georgia would allow me to establish my academic credentials while taking us on a family escapade. After a couple of years, as jobs opened, we would return to Canada, or so we thought. I signed on the dotted line.

Life has a way of writing its own script, but for a while there, we actually believed we knew what we were doing

I Need Four Seasons

Our first impressions of the South were not terribly good. We arrived in the heat of late July in 1980. I saw my very pregnant wife looking out the car window and I knew she was thinking, "Where did he bring me?" I, too, could not believe that man nor beast, and certainly no herbaceous plants, could possibly live here. Housing was inexpensive in Athens, and we purchased a lovely home on the east side on a two-acre lot with dozens of mature white and red oaks. Essentially, we lived in the forest. This was so different from what I was used to; even in lovely East Lansing, large lots were not at all common, and the treed lots so prevalent all over Athens just did not exist there.

The first spring in the South was unlike anything I had experienced.

Dogwoods, azaleas, rhododendrons, peaches, pears, magnolias, and wisteria filled the senses. Even the yards of people who did not care about gardening had some redbuds or dogwoods growing wild. I was in awe. I gushed to friends back home about that first spring, and to this day, spring is still a brilliant tapestry; a palette that blossoms forth with little apparent effort. This might not be such a bad place after all.

Athens is in the northern part of the state and geographically in the foothills of the Smoky Mountains, so it has the same four seasons as Quebec and Michigan. Being from so far north, I looked at the seasons differently from the locals. The biggest difference between my perception of seasons and that of the locals was when seasons started and finished. To me, spring started in February and lasted through until June. Fall began in October and went through to December, leaving only a couple of months of winter. When an inch of snow fell, everyone attacked the milk and bread shelves at the grocery store—I considered it a wonderful and short diversion. When a serious frost visited, people bundled up—I loved it. I missed the snow and cold, but I didn't miss five months of it.

Summers were and still are a bear. However, like anything else, once you learn to work with it rather than against it, it is not so rough. In the heat of the summer, we worked outside in the mornings and inside later in the day. We went jogging in the morning; the kids' soccer and baseball games were scheduled to end by early afternoon. The weather was different enough, but so many things in the South took some getting used to, especially by a foreigner. I still cannot do grits, collards, or okra.

Music, theater, and restaurants kept getting better over time. Being a university town, Athens was diverse, attracting people to the university from all over the world. Our neighbor to one side was from England; the other side, Korea. My colleagues at work were from all over the country and all over the world. Our friends and acquaintances were equally diverse. In 1980, it was a small Southern town with a good university, and somewhat sleepy. Today it is big small town with an excellent university, and too busy to sleep. The more I travelled, the more obvious that became. In short, while Athens, Georgia, was not Montreal, we were very much getting used to our new surroundings.

School Days

One of the most important considerations in any move is the kids' schools. We looked at the public schools and were more than pleased, so within a month of our arriving, we took the girls to meet their teachers.

. . . .

Heather was always adaptable to change. Whoever penned the saying, "Go with the flow" must have had Heather in mind. Being significantly taller than other children, Laura took many months to feel comfortable in new situations. When we told them that we were moving from Michigan to Georgia, Laura hated the idea. She was eight years old. The thought of attending a new school terrified her.

The secretary at the school pronounced her name with a southern accent, slowly and with two long drawn out syllables. When Laura tried to correct her, this fine Southern lady simply said, "Don't worry, honey, you will be pronouncing it just like us by the end of the year." She was right. Laura had never taken a school bus before and was not looking forward to such an unknown experience. Heather, who was starting kindergarten, couldn't wait to get on board.

We drove the two kids the first day. Heather ran off with a quick kiss to find her class, but Laura squeezed Dad's hand tightly and fought off tears every step towards the classroom. Once she was seated in her class, we parted with the kind of heavy hearts that only parents who have left their kids "for their own good" can know. Heather was bubbling after her first day, telling us about her classmates, teacher, and their funny way of talking. Laura's silence, however, said far more than Heather's words. Guilt has a way of rearing its ugly head when things go badly. I had felt awful enough for moving the family further from their Canadian roots, and in particular, exposing Laura to such trauma. Because the weeks prior to school had passed smoothly, and friendships were starting in the new neighborhood, I had stored up a little false optimism that school might not be too bad.

Laura would not go to the bus by herself, hated the idea of going off to school, and could make no new friends. After three days of this, her sister said, "Don't worry, Dad, I've been taking care of her." Sure enough, Heather, who would have loved to run off to the bus and skip merrily to her new classroom, walked slowly with Laura to the bus stop, and then escorted her to her class. This seems like a small thing, but it was a gesture of love on Heather's part. Every day, for about four weeks, a small five year old walked her older and taller sister to her class. Laura never asked, but there seemed to be an unspoken agreement between the two.

. . . .

One should not get the impression that this type of cooperation was at all normal. They usually fought like cats and dogs, and the term sibling rivalry was one that they helped popularize. Soon enough, although she still hated school, Laura showed the same determination she had with the bottle. Shoulders back (no more slouching to look short), lips slightly bitten, and without complaint, she took her place each day. It took about six months before she was comfortable, but one of the reasons she persevered was because of the quiet, unasked-for help of her little sister.

Crop Chats with Dr. A.

I soon became immersed in the duties of an assistant professor. My experiences in high school had prepared me to work with students, but there were huge differences between university teaching and a high school classroom. I was to teach a class in greenhouse management and another in greenhouse crop production. At the university level, there are no state guidelines or state testing; the entire course content is entirely up to the department—in this case, me. I decided they should learn greenhouse systems, such as heating and cooling, environmental inputs

Amy Gard'ner

like photoperiod, lighting, and temperature. Most importantly, I wanted them to experience hands-on growing of many crops.

I quickly noted that the kids in my classes were far better students than I ever had been. Not all my horticulture students were brain surgeons, but they were bright. Most of them had discovered horticulture after enrolling in other options. Nearly all were juniors and seniors and were in my classes because they wanted to be, not because they had to be.

We had a ball growing plants. Over the years, we messed up spring bulbs, Easter cacti, all sorts of bedding plants, and hundreds of lilies. If it is true that one learns most from one's mistakes, we were brilliant. The most difficult were poinsettias, but we manipulated photoperiods, nutrition, and growth regulators. And oh my, we produced some of the ugliest plants known to man. The easiest and most rewarding were pot mums, because we could schedule using day length, we could pinch - or not, and they were sufficiently forgiving that we were able to produce reasonable crops. The number of gloxinias tossed out was legion, and the student stress over Easter lilies was palpable.

I was not teaching stress. I was trying to instill responsibility. The students were responsible for absolutely everything concerning their crops. They were in charge of soil choices, potting, watering, nutritional needs, growth regulators, disease and insect recognition, pinching, and disbudding. If the student was out the night before and did not water his or her plants the next morning, no one did it for him or her. Every day, weekends and holidays included, the plants became their life.

Every crop had to be scheduled. Poinsettias for early December and lilies for a week before Easter caused immense stress for these would-be growers. For most kids, this was the first time they had been in a greenhouse, so I was not expecting picture-perfect crops. However, I expected them to know why they did, or did not, water, pinch, growth regulate, or feed their plants. It sounds all quite gruesome, but I never had a problem filling the classes. However, if there was one particularly frightening aspect of these classes, it was likely the dreaded Crop Chat.

At the end of the semester, each student and I would spend thirty minutes in the greenhouse with all the crops he or she had grown and we would "chat." Oral examinations seemed to have disappeared from undergraduate curricula in the 1960s, perhaps because educators believed them to be forms of cruel and unusual punishment. For some reason, my students did not consider being interrogated an occasion for learning, but it was. The Crop Chat became part of student folklore, and when Crop Chat Day came around, every professor and student in the department knew about it. To be sure, creativity was alive and well. The reasons provided for stunted plants, misscheduled plants, or dead plants were ingenious and often hilarious. In spite of myself, sometimes I could not help but break out into laughter. In such a setting, it was obvious who paid attention to detail and who simply went through the motions. Unfortunately, a number of students left the greenhouse crying, but that was not my intention. I felt bad, but oh my, how the next ones were prepared.

I was involved in teaching the greenhouse classes for over twenty years. A few students became tired of me, and me them, but we all parted as colleagues and friends. So many of my students have succeeded in this industry because they were given a chance to make mistakes.

Brian Dempsey, Cody Gambrell,
and Kyle Joiner

Over those two decades, the soil, pots, flats, fertilizers, equipment, and plants used in these courses would have cost the department tens of thousands of dollars. They cost nothing. The support from dozens of companies allowed students the hands-on experiences that could never be taught in a classroom setting. Such cooperation between industry and academia is unknown in many countries, but it encourages bright young people to join our industry. In return, we are richer because of this teamwork.

Learning the Difference Between a Peony and a Petunia

In the late 1980s, no one was teaching students anything about herbaceous plants in the landscape. Woody plant identification was a mainstay of all horticulture programs, but in many universities, herbaceous plant identification was of little concern. This was the time when annuals were breaking away from the bedding plant moniker.

Unknown plants like scaevola, bacopa, and calibrachoa, were bursting onto the landscape scene. A new company called Proven Winners® introduced its own brand of plants and seriously shook up the traditional plant business.

Shanna Tomlinson, taking notes in Perennial Plant ID class

Large companies like the Ecke Corporation®, Ball Horticulture®, and others soon joined in the pursuit of new labels. And swept along into this tsunami of new annuals was an old but static plant group, herbaceous perennials. With the emphasis on bedding plants, perennials had fallen out of favor and had not significantly changed since the 1940s. Students were graduating from horticulture not knowing a peony from a petunia. In 1986, I offered to teach a class on perennial plant ID and soon added another class to expose students to the new annuals as well. These classes became very popular. Although most of the students probably forgot eighty percent of the names after a couple of years, they were certainly better horticulturists.

One of the best parts of teaching the perennial class was the community involvement. Most departments do not have sufficiently large gardens, or the timing of the classes simply doesn't follow the natural seasons. It may be too cold to study them outside. Fortunately, in Athens, spring started sufficiently early, and fall stayed long enough that I could use the outdoors as my laboratory. The trial garden was barely up and running, and I quickly realized that students would learn far more if we could conduct classes in real gardens, rather than with images on a screen. I asked local gardeners if they would open up their properties for students to learn. They did so with gusto. On Wednesdays, we would meet at various gardens in town and identify and talk about the use and value of dozens of perennials—in a real garden. As wonderful as the plants were, the gardeners themselves were even better. Students were absolutely charmed by the passion of the gardeners and the stories of their garden's evolution. The plants themselves and the tea and cookies didn't hurt either.

Later, the kids would bring their parents, roommates, and girlfriends/boyfriends back to the gardens, and wonderful friendships evolved. The gardeners loved being included, not only for the interaction with these young people, but also for the knowledge they picked

Gardeners Donna and Ed Lambert, a fabulous couple who inspired students every year.

up during class. As one lady said, "This is so wonderful! Now I finally know what is in my garden."

Publish or Perish!

Promotion and tenure (P&T) are major stepping stones in the career of any professor. If the standards for P&T are not met, the professor is fired, usually after five to seven years. In major research universities like UGA, those standards revolve around research publications. Very little else is important; poor teaching evaluations are given little weight, and unless the professor is a serial killer, an exemplary research record will get him or her promoted. Every young professor knows this, so it is not surprising that grant writing and time spent on research are far more important than time spent in the classroom. I was certainly no different.

I diligently researched environmental influences on plant growth in the greenhouse and managed to publish my findings in refereed publications. There are few things more frightening than submitting a publication for peer review. Once submitted, it will be read by at least three other professors at other institutions, each one likely knowing more than you about the subject. When the reviews came back, I often waited for three days or more before I had the courage to open the envelope. Most times the outcome was "accepted, but with changes" and occasionally outright rejection. It is a time-tested and objective method of maintaining high research standards. It is nevertheless extremely stressful. Even when I had published dozens of papers, the feedback from my peers was harrowing. "Publish or

Lily research with Robyn Bellis

Perish" was a well-known and accurate axiom of university life. I had no intention of perishing.

I Stole It

At Michigan State University, Dr. Carlson had created a wonderful trial garden to assess landscape performance of annuals. It was a source of pride for the department and was used for teaching, extension, and research as well. Will told me, "The best thing about the trials is that we can interact with the breeders and industry in a very tangible way." Not only were the trials a highly visible and pragmatic example of research, they were supported financially by breeders throughout the world. Growers, breeders, landscapers, and garden center operators would come by and take notes about plants that looked good and order accordingly. There was nothing like this in Georgia, and I believed this should be changed. So, in 1982, I talked to one of my colleagues about starting a trial garden.

One of the best things about the Horticulture Program at UGA was that we were not micro-managed. We all knew the "rules," but no one was hovering over us telling us to spend more time on this, or less time on that. We had yearly reviews, and if we were not getting the job done—essentially meaning if we were not publishing enough—the department chair would encourage us to spend more effort and time for research. However, if you remained on track, few questions were asked.

In 1979, Dr. Michael Dirr was recruited from the University of Illinois to jump-start the State Botanical Garden in Athens. His passion for plants and his textbook, *Manual of Woody Landscape Plants*, had already made him a superstar. We met the next year and hit it off immediately. I helped out at the botanical garden when a large project was due and he needed a little extra help. He raised significant funds for a new conservatory, added new gardens, and raised the consciousness of the Botanical Garden.

Nothing was easy. He was forever fighting for a larger budget and haranguing university administrators every chance he got. Michael was a hands-on person and absolutely unsuited for the wasted time of university politics. After arguing one more time for money to keep the lights on, he threw up his hands in disgust. He was ready to take himself, his wife Bonnie, and their three kids back to Illinois. The university asked the department if we could find a position for him, and after a discussion of at least a minute and a half, Dr. Michael Dirr was a professor of horticulture. Team Armitage and Dirr went to work.

Allan and Michael Dirr, 2013

Michael is very opinionated, does not suffer fools lightly, and is sometimes hard to love. He put up with me, and me him, and together we did our part in helping elevate the reputation of ornamental horticulture at UGA. No other person helped me in my career as much as Michael. He encouraged me to write a book, he invited me to join him on garden tours, and he suggested taking students on extended field trips. If I ever got a swollen head, I just had to look at Michael's fans and followers to know that I could never be the star he was. If I ever thought I knew every plant out there, I just had to watch him stare at a bud

Trial Garden, 1984

on a tree anywhere in the world, take off his glasses, stare again, and then tell everyone what it was, often down to cultivar. I have yet to meet anyone with near the expertise in woody plant identification as Michael. If he had not been there, I doubt I would have accomplished half the things I did. But in those early days, he was just a big doofus like me, so I pulled him aside and asked, "Hey Mike, want to start a trial garden?"

Rather than tell you tales about how many university committees I had to go through, or how many hoops I had to jump through to procure property for a garden, I'll come clean. I stole it.

An unused piece of land lay just behind our greenhouses surrounded by other buildings. It seemed to me that nobody would mind if we upgraded it. The saying, "Better to ask forgiveness than permission," made absolute sense in this case.

We volunteered a few students and built a small arbor, dug a few beds, and put some plants in. Mike recruited woody people like Don Shadow from Tennessee to donate special trees and shrubs. I approached PanAmerican Seed® in West Chicago and Harris Moran Seed Company® in Rochester, New York, and they immediately jumped on board. Expanding the plantings, providing data to those few breeders who were paying attention, and finding students to work was not without its headaches. However, the garden was small enough that I could grow the seed material in my class and find students to help maintain it. Michael's help in getting the garden going was invaluable, but in a year or so he saw his colleague taking over the space with "wimpy" annuals. He moved his trials to the much larger horticulture farm.

The garden was very much a barometer of new plant breeding. When it was started, about ninety percent of the annuals were seed-grown, ten percent vegetative. Today it is the other way around. We only trialed annuals because that was all that large breeders were doing. Small perennial breeders had little interest in sending out material.

UGA Trial Gardens, 2013

Honorary plaque to Dr. A, UGA Trial Gardens

By the early 1990s, the explosion of vegetative annuals, which we coined "specialty annuals," occurred. More and more plant breeders, including those working on herbaceous perennials, joined us. The garden's mission was research then and still is today. We published data and distributed the results to anyone participating. It was starting to look a little like a real trial garden.

The garden was proving to be a godsend. Not only was it attracting breeders from around the world, but also local growers, landscapers, and, of course, gardeners. All breeders paid a small stipend to have their cultivars in the garden. I did not have to go to the department or the university with my hand out. Added to those benefits, I made sure there was always a diversity of plants to function as a teaching facility as well. You would think that the university would have been proud.

Instead, almost every other year, the administration eyed that property as a potential site for the expansion of existing buildings, especially the pharmacy building, to which it abutted.

My patient garden manager Meg Green and I would chain ourselves to the mature trees Michael had originally planted. We wrote letters and obtained signatures to stave off annihilation until the next time. We kept them away. We even managed to accomplish some major renovations, so that it would be harder to annex the next time. In time, the garden became quite well known and was cited many times as one of the five best trial gardens in the country. It is still there.

I hired dozens of extraordinary young men and women over the years to work in the gardens. It was a highly sought-after job, and I was able to pick and choose the very best from my classes. I am most proud of this aspect of the garden.

The worker bees, 2012 : Meg Green, Blaine Pritchett, Erik Durden, Johanna Biang, Robby Jourdan, Holly Hess, Candler Woods, Brett McAdams

It helped teach not only the entire cycle of propagation, greenhouse production, planting, maintenance, and data collection, but, much more important, the meaning of responsibility. They showed up on time, worked between classes, sweated in the ninety-nine-degree heat, stressed out together when open houses were scheduled, and learned cooperation and pride. When breeders, landscapers, garden center owners, and gardeners descended on the garden, their compliments helped the students realize that they were making a difference. When companies asked about potential employees, I had an entire stable of thoroughbreds. Nearly every one of the students who worked for me went on to very successful careers.

Nine

"What no wife of a writer can ever understand is that a writer is working when he's staring out of the window."

. . . .

- Burton Rascoe

Big Jon

Meanwhile, the family was getting the hang of this Georgia thing. Yes, it was hot in the summer, and we would likely never get used to the upside down feeling of being inside on a summer day and outside in the winter. After a couple of years, we were a typical nuclear family. We had a dog, a house, some property, and our third child had arrived. Susan likes to call Jonathan our caboose.

Jon was born in October 1980, seven years after Laura. He was a big boy, and we often called him Big Jon.

At first, the girls were disappointed that he was not a girl, but they had no trouble dressing him in high heels, skirts, and bonnets along with gaudy lipstick and rouge. Jon's arrival was like having a second family. Because he was so much younger, there were few conflicts in sporting events or school plays with the girls. That is the way we are with our children—we micro-manage the eldest to the point where we come close to messing them up. The next one is a little easier, but we are still conflicted with providing enough time for each. Fortunately, we become more laid-back with each subsequent child. With Jon, we had far less angst and, as a result, he was almost stress-free. That is not to say he was not a handful. Sometimes he filled his hands with things he should not.

. . . .

Jon was not a bad kid; he simply found himself in situations where trouble was his closest companion. The girls had taught us that certain situations should be carefully avoided whenever possible, but with Jon, they appeared as if by spontaneous generation. On this particular day, things were progressing smoothly, which is to say that Jon had not yet broken anything or gotten into any serious trouble.

A few miles from the house was a small old mill that sold children's clothes. Today, it has gone the way of mills everywhere, but at the time, clothes were offered at bargain prices through the outlet store, equally small and aged. This sale occurred but twice a year, and on such days, every woman in the entire state of Georgia knew of the sale. It was a point of honor to be there at opening.

Women were poring over the clothes in a search and destroy mission that would have made the Pentagon proud. Upon looking back, it seems obvious that this was no place for a two year old, but the bargains were too good to pass up. For the first fifteen minutes, Jon was the golden boy, but as boredom set in, he started to toss about some of the clothes on the tables. When Susan said, "No" for the fourth time, Jon decided to have a temper tantrum. Imagine this large small boy on an old wooden floor crying and yelling something incomprehensible, while waves of women stepped over him jostling for position at the clothes table. Since Susan was paying little attention to him, he decided to use his ultimate attention grabber. He calmly reached up and pinched Susan on the bottom, hard. This tactic had been successful before. It likely would have been again, except for one problem. It wasn't Susan's bottom.

Above the noise and the din, a piercing scream was noted by one and all. Silence fell. Big Jon had applied one of his finest efforts to the ample bottom of the woman nearby, who jumped to heights never again attainable in her lifetime. The whole place became still; even Jon was quiet, as he figured something was not quite right. Susan was mortified. The woman was angry, embarrassed, and more than a little sore. If looks could kill, my wife and son would have perished that day. Susan quickly picked up the kid and beat a hasty retreat to the car. She bought no shirts or pants that day; Jon's clothes were purchased in the safety of the mall. Not as inexpensive to the pocketbook, but a whole lot easier on the mind. Susan seldom went back to the mill store, probably in the fear she would run into a woman with a large, scarred backside. One good thing did come of the adventure: Jonathan seldom pinched after that. It turned out that Jon's education was always to be littered with great doses of creativity.

. . . .

Throughout my career, family balanced everything. Not that I was the greatest husband or dad in the entire universe, but the escapades of the kids growing up always put things in perspective. Our children had no more misadventures than anyone else's; I simply wrote them down. As I look back on some of those times and see the children where they are now, I can do nothing but smile at our good fortune.

Young and Inexpensive

During my first few years at UGA, I put my head down and tried to do all the things necessary to support the family and to be successful. I taught, I wrote, and every now and then was invited to speak at state and national meetings.

In 1983, I was just finishing a tree house for my daughters when the phone rang. It was Paul Ecke, Jr. Mr. Ecke was a Godlike figure to everybody in the industry. People associated the Ecke name with poinsettias, but he was also an admired mentor to young people. His scholarships and research foundation funded many young scientists, and quite simply, he was a nice man. That he would be calling me on the weekend was entirely unexpected. Even more unexpected was that he was offering me a job in his upper management team at the Ecke Ranch in Encinitas, California. We had just arrived in Athens, and another move was the last thing on our minds. I was happy at UGA and not looking for anything else. So, as flattered as I was, I thanked him and said no.

Every week for at least six weeks, magnificent coffee table books titled *Beaches of California*, *Mountains of California*, or *Vineyards of California* arrived in the mail. When friends visited, they were most impressed with our love of California. When two airplane tickets arrived with an invitation to visit, we could not say no. The coast was gorgeous, the accommodations first class, and the hospitality gracious. I toured the ranch and met the people. It was a remarkable place.

When we were returning home, we talked about the opportunity, but we both felt that we were just getting started. The kids had been uprooted too many times, and I believed I would miss the writing and student interactions of a university. That the house prices were astronomical entered into our decision, but only to a slight degree. The next day I thanked him and declined the offer. In the next five years, many opportunities came my way. After all, that was the time when I was young and inexpensive. Michigan State, University of Guelph, and a number of private sector jobs arose, but we were having too much fun where we were. So we stayed at UGA, and then they couldn't get rid of me.

The Six-Month Writing Gig

I found myself surrounded by a constantly changing industry. I wanted to be part of it. The 1980s brought important developments in automation, greenhouse construction, the advent of plugs, advances in nutrition and growth regulators, and the emergence of box stores. Think of any of those topics today; they have all been part of industry background for thirty-plus years, but then they were mere shadows of what they are today. Industries always change. Ours is no different;

First issue, Greenhouse Grower, *January 1983*

however, it seemed I had jumped in a maelstrom of extraordinary times and people.

One such person was Dick Meister. He is an amazing man, one with vision and foresight, and willing to take a risk. He and his brother Ed had expanded their father's publication, *American Fruit Grower*, in the 1940s. With Will Carlson's encouragement, Dick launched a new magazine for professionals and academics in the floriculture industry. Even though the granddaddy of publications, *Grower Talks*, had been established in 1937, Dick and Will believed the industry would embrace this new publication as well. In January 1983, the first edition of *Greenhouse Grower* magazine was published.

In 1984, I was asked to write a column. I jumped at the opportunity. If I showed some ability to write, and reviews were positive, I was told I could continue for six months. My first column appeared in April 1984, titled "Perennials Have Potential." While it did not win any prizes for literature, the information was solid, and I did not get fired. The next month I wrote about acclimatizing bedding plants for retail sales. I was able to share my research with industry people who would otherwise not read it in a scientific publication. I wrote about plant trials, new introductions, and the world as I saw it. I felt very close to my audience and wrote as if I were talking to them about something that could change their lives.

It was thrilling to sit down and pen something each month, stimulating to

Dick Meister, Greenhouse Grower

share ideas, and utterly terrifying. All writers will tell you in unguarded moments that a writer is naked once his words are printed. People can take potshots; they will disagree and argue, often quite vehemently, with what you have written. It is far easier to get along in life by doing the minimum than by exposing your passion each month. However, ignorance is bliss, and I approached each column that year with fervor. I truly was afraid I would soon get the pink slip.

Six months passed, and then a year, and I was still at it. No sooner had I looked up than I had been writing the column for five years. As I sit here in 2015, I am thinking about what I shall share next month. I have written about travels, philosophy, and penned thousands of words about plants.

Oh, have I written about plants! When vegetative annuals broke onto the scene, I called them "specialty annuals" and described uses and production schedules. Nothing was safe from my pen: cut flowers, perennials, and even my daughters were fair game.

I provided my opinion, I told my stories, and I am still hoping not to get fired. I have worked with fabulous lead editors such as Jane Lieberth, Robyn Dill, Delilah Onofrey, Kevin Yanik, Robin Siktberg, and now Laura Drotleff. I am the old man in the company, but they still put up with me. It is a great organization and continues to be a great ride.

Columns have changed, to be sure. I incorporate much more about marketing, but I still talk about how my daughters view the things we do. I probably tell too many stories but always fall back to new plants.

When I read old columns, I see genera and cultivars that were just emerging then but are commonplace now. I read about problems I wrote about then that are still problems today. Someday someone might compile *The Selected Columns of Dr. A* and put some of the better ones in a book of its own. You never know: it is thirty-one years of history, without the politics and wars.

Ten

"Don't send me flowers when I'm dead.
If you like me, send them while I'm alive."

. . . .
- Brian Clough

Mama, Someone's Here About Your Flowers

The more breeders I met and the more new plants I saw, the more I understood what made a successful plant in the marketplace. Introductions of bedding plants have always been most common. They are relatively easy to breed, the market understands how to use them, and the buyer is comfortable with them. In the late 1980s and early 1990s, few genuinely new plants were being introduced, but times were changing. To succeed, a new introduction had to be easy to propagate and grow well in a container. In theory, it had to look good at retail, be inexpensive to grow, and do well in the landscape. The reality, however, was that if it was producible, pretty, and inexpensive, it would be introduced. If it also did well in the landscape, that was a bonus.

Having the trial garden allowed me to evaluate the landscape side of the equation, and while nobody might listen, I could call a dog a dog. Athens was not Chicago or Philadelphia, but if plants did well in our growing conditions, I would write about them in *Greenhouse Grower* and shout them out on the lecture circuit. If they did poorly, I could do the same. It was also becoming apparent that large migrations of people were moving south to cities like Atlanta, Charlotte, and Dallas. Heat tolerance became a catchphrase.

Breeders actually started to pay attention. All these things were swirling about when Mike Dirr and I drove to a meeting in Atlanta in early March 1990.

. . . .

"Look," we both said at the same time. "What is that?" The "that" was a patch of purple in a yard across the highway. Without hesitation, Mike turned his little red truck around, and we headed over to investigate. It turned out the purple patch was in front of a small house, replete with dead cars, washing machine on the porch, and peeling paint. In short, it was a house you drove by, not to. We slowly coasted down the dirt drive with more than a bit of trepidation. I must admit that the banjo music from Deliverance *played in the back of my mind. Michael was bigger than I, so I put him in front. When he knocked on the door, a lanky, bare-chested teenager opened the door. "Would you mind if we looked at the flowers?" He eyed us quizzically, then turned around and shouted, "Mama, someone's here about your flowers."*

Verbena 'Homestead Purple'

Mama soon came out, joined by daughter, cats, dogs, and other assorted two- and four-legged critters. We asked her if we could look at her plants. Her face broke into the widest grin—I believe that was the first time anyone had ever complimented her on her flowers. We walked over en masse and I exclaimed, "Mike, these are verbena!" No sooner were the words out of my mouth than I received a staggering elbow to the side. "Son," she looked at me and said, "They ain't verbena, they be vervain." She was using the old-fashioned name for verbena. I was properly chastised. I had seen a lot of verbena but never one so vibrant or so early. It also possessed significant cold tolerance, as it obviously had come through many a Zone 7 winter. I asked her if she knew where they came from. "Nope, but they've been here for years." When I asked if she would let us have some cuttings, her smile was still evident, and she nodded yes. We took seven cuttings with us that day and put them in the trial garden in April. In recognition of the dwelling where the flowers were found, we quickly came up with a name—Verbena 'Homestead Purple'.

. . . .

I presented a number of new plants to the industry each year that I believed to have some potential for sales. These would include things I had found or selected. They may have been common plants like a new impatiens, less well-known ones like *Acalypha*, or plants no one had heard of like *Tibouchina*. I even showed off a new plant called *Angelonia*. Most were a little too far out at the time, but in all cases, I allowed growers to take cuttings free of charge. If they were sold, I simply asked them to keep the name I had given and tell people it came from UGA. Cuttings were taken for many plants I put out, and some were even moderately successful. 'Homestead Purple' was planted in the trial garden in the spring of 1990. Growers immediately took cuttings. By 1992, it was sold throughout the South; by 1994, it was the number one selling verbena in the marketplace and quickly made its way to Europe as well. Even today, it still sells in the millions of units. It has been the standard of verbena for years and is still used as a parent for new cultivars.

In 1990, universities did not patent plants. University administrators were absolutely clueless about the numbers of ornamental plants sold. Even if they were not, it was academic policy not to "sell" research; after all, we were a public institution relying on tax dollars. I had no problem with this policy. I was simply happy to see the fruits of my work appearing in landscapes here and there. This policy quickly turned around as the next decade approached, but 'Homestead Purple' never made a penny for Michael, the university, or me.

About five years after 'Homestead Purple' was introduced, I went back to find the woman and thank her. Her land had been sold. The house had been leveled and replaced with a strip mall and a parking lot. I don't know when that occurred, but I do know that if we hadn't driven down the driveway and knocked on the door, a wonderful plant would never have been introduced. One thing's for sure, I will always remember her face lighting up, all because of a little patch of purple on a cold March day.

They Don't Grow in Full Sun

The trial gardens were fulfilling all my expectations. Breeders were sending plants, companies were visiting, landscapers, and gardeners were taking notes. I had my own new-plant lab and never had a problem introducing new plants wherever I spoke. We published the data so its main reason for being—that of a research facility—stayed true. Annuals, especially bedding plants, were the underpinning of the greenhouse industry, but in the late 1980s and early 1990s, perennials were gathering momentum. However, to "mainstream" them, breeders and distributors had to be involved. Up to this point, perennials were mainly sold to enthusiasts through mail order. As more were introduced and found their way to garden center racks, both the retail and mail order trades grew exponentially. Compared to what we see today, the perennial market was merely an outline of what they were to become. I had been trialing perennials since the beginning of the garden, but expanded the perennial trialing in the early 1990s. By 1995, perennials covered the entire perimeter of the garden. Soon, I was awash in an explosion of new introductions of *Heuchera*, *Pulmonaria*, and *Achillea*.

Coleus 'Rustic Orange'

My, it was an exciting time!

Our success with new plants, especially 'Homestead Purple', caught the eye of colleagues, and people brought in all sorts of plants they discovered in their backyards. I never had the patience to be a "real" breeder but I knew enough to see the potential of unknown plants to the marketplace. I call it a "horticultural eye," the ability to see something unique that thousands of others have walked by. This eye allows breeders to cull their seedlings and decide which one of the hundreds of possibilities will actually reach the market. All great plantsmen and plantswomen have that horticultural eye, and if they come across an interesting plant not in the marketplace, they will investigate further.

Such was the case with 'Homestead Purple', and such were the eyeballs of my friend and colleague, Dr. Butch Ferree. Butch was a well-traveled extension agent who arrived in July with a bag of cuttings under his arm. I am always excited to see plants, but I must have shown my disappointment when I looked inside and saw some tired-looking coleus. "These are special," Butch said. "They were growing in full sun!" "Butch," I said, "Coleus don't grow in full sun."

He had brought fifteen cuttings, each a different color and said, "Try them out, Allan, they truly were growing under the blazing sun in Louisiana." The more I thought about it, the more the idea intrigued me. Plants rooted in less than ten days and were planted in the sunniest part of the trial garden in late July. I kept waiting for the poor plants to burn up, but the heat of August and September did not faze them. I was impressed. I propagated many of them over the winter and put ten of the fifteen plants in the trial garden in spring of 1994, calling them the Sunlover series.

With the help of Dr. Ferree and an excellent grower, Mr. Sammy Turner, we named seven of the new clones. During the next open house, I introduced our visitors to the Sunlover series consisting of 'Thumbellina', 'Rustic Orange', 'Gay's Delight', 'Red Ruffles', 'Freckles', 'Cranberry Salad', and 'Olympic Torch'. The only sun-tolerant coleus then was called 'Bellingrath', named for the famous gardens in Mobile, Alabama. It later became known as 'Alabama Sunset'. Although 'Alabama Sunset' was on the market, it was not well known outside the South. In fact, coleus for full sun were not known at all. That soon changed.

The Sunlover series was the first of sun-tolerant coleus. It was an instant hit. Within a year, many of the cultivars were sold throughout the United States; within three years they became popular everywhere. It took no time before breeders were hybridizing and selecting coleus for full sun.

The first to follow up was my friend George Griffith of Hatchett Creek Farms in Gainesville, Florida. He introduced the Solar series the next year. They were truly extraordinary. With that, the tidal bore of sun-tolerant coleus grew, and continues yet today. Over two hundred coleus can be found in the marketplace today; yet, twenty years later, I still see 'Rustic Orange' everywhere. 'Gay's Delight' and 'Red Ruffles' are also favorites, although often under different names. Even though none of the plants were patented, they managed to change the face of landscaping to this day.

They Laughed Politely

As I look back on my involvement with this thing called new plants, it is obvious that that topic was accelerating nationally while I was a young man at UGA. The Southeast was populated by brilliant people like Don Shadow, Tony Avent, Ted Stevens, and Ray Bracken. Michael Dirr's love of woody plants and his passion for the unusual were contagious. Between the two of us, it was hard to get a word in edgewise.

With so many excellent plantspeople in the Southeast, it became a mecca for new plant research. One of the best things we did was organize the Southeast Plants Conference in 1989.

We invited independent breeders, mostly woody plant people in those days, to present what they were working on and where they saw the market going. Kodachrome slides (no PowerPoint then) flashed on the screen. The depth of diversity was amazing. However, out of the ten or so speakers, only two were doing anything in herbaceous plants. I talked about some of the new annuals I was looking at and spent time talking about an up-and-coming group of plants—perennials—and how I believed they were poised to explode. The gathering was such a success we decided to meet two years later in Raleigh, North Carolina.

The meeting in 1991 attracted well over one hundred growers, landscapers, and researchers, and more than a dozen speakers. Additional speakers appeared on herbaceous material, but the meeting was still dominated by woody plants.

When my turn came, the moderator introduced me as the fellow who plays with "those wimpy plants." I talked about the many perennials we were looking at that were building up steam. Then I changed tack and showed a photo of a black-leaved plant I had been given a year earlier. A volunteer had rescued it from a trash bin at the USDA facility in Monroe, Louisiana. I received a single cutting, planted it, and watched it grow. The plant was a sweet potato.

I told the audience it should be called 'Blackie.' I suggested that this might make an excellent ornamental. They laughed—politely of course—after all, this was the South. "Who in their right mind would think you can make money selling a sweet potato?" was an oft-expressed question at the meeting. I held my ground, as I believed its ease of propagation, rapid growth, landscape color, and heat tolerance were positive attributes. A few people asked for cuttings, but a discussion of new dogwoods was on the docket next, and that was that.

At the end of the conference, a young man approached me and said, "Dr. Armitage, I can't believe you talked about sweet potatoes, but I have one I would like you to look at." His name was Hunter Stubbs. He brought me a sickly looking plant with two small chartreuse leaves, struggling in a four-inch pot. I was not sure it would even live through the car ride home.

The first planting of 'Margarita' sweet potato, the Trial Gardens at UGA, 1992

I said, "Hunter, just in case this plant is any good, what would you like to call it?" He thought for a while and then said, "How about 'Margarita'?" I put the plant in the garden, and its vigor and brilliant color surprised all of us.

During open house the following season, everyone wanted a piece.

Within two years, 'Margarita' was grown throughout the South; within five years, it was everywhere; and in a few more years, it was impossible to walk by a container anywhere in the country without tripping over it.

Since then, at least a dozen cultivars of sweet potato have come on the market. They may have chartreuse, purple, or variegated leaves. They can be found with heart-shaped to cut leaves and with different vigor. However, twenty-five years since that meeting, 'Blackie' and 'Margarita' are still the most popular sweet potatoes in landscapes today. Every time my daughter passes a 'Margarita' in a container or a garden, she still says, "Dad, that could have been my kid's college tuition."

That's All I Recognized

As we introduced more plants at UGA, people started thinking of the department as a new plant magnet. There was no way I could, nor wanted to, compete with breeding companies. They employed dozens of breeders, hundreds of sales and distribution people, and had state-of-the-art facilities. My program consisted of a run-down greenhouse, circa 1940, two greenhouse staff, a few enthusiastic students, and a trial garden. However, there are many ways to introduce new plants—I was having success using non-traditional breeding techniques. I believed many of the plants I was touting had market potential based on my "horticultural eyeball." It had been successful with plants I had "discovered," as well as with one or two that people brought to the program. I was having a ball.

In the late 1980s, one of the greenhouses had some random odds-and-ends plants, as all greenhouses tend to collect. But most of it was filled with my new discoveries. It was overflowing with dozens of bright, wild, and exciting flowers. They were spilling out of baskets, tucked in beside other new plants, in oranges, reds, blues and yellows. Three-feet-tall perennials competed for attention with stunning, dwarf, compact annuals. When plantspeople walked in, their jaws dropped. I had no doubt that I was one of the finest breeders of my time.

. . . .

At the end of a particularly tiring day, Carl Lasco, the greenhouse manager, and I were relaxing in the office just off the greenhouse. Around five p.m., a lady, calling herself Rachel, walked in and inquired if this was the Hort Club plant sale. I said that these were research facilities and the sale was elsewhere. She asked if she could walk through the greenhouses, which she did.

About fifteen minutes later, she came into the office carrying two pots. She had her purse open, money ready, and wanted to know how much they were. I was about to repeat my former statement when she put the two pots on the desk. In front of me, she placed a four-inch pot of 'Better Boy' tomato and a six-inch pot of Leyland cypress. I was astonished. I looked up and said, "Rachel, when you walked in there, you believed that everything was for sale. You were surrounded with incredible beauty and sensational plants. There were no prices listed, you could have chosen anything at all." She nodded in agreement. "Then why did you choose a four-inch pot of a 'Better Boy' tomato and a six-inch pot of Leyland cypress?"

At that, she looked me in the eye and said, "That's all I recognized."

. . . .

No longer was I astonished. I was deflated. Then I understood. I looked at new plants the way a plantsman would see them. I was judging them the way a salesperson would judge them. In fact, I was looking at them every way except the way my daughters would look at them. What a lesson I learned that day! It is easy to get carried away with "new," but it is even easier to forget who the "new" is for.

Comfort is sometimes more important than discovery. It took me a while to appreciate Rachel's visit, but comfort of the buyer became an important feature of my future plant research. By the way, Rachel was a happy lady when she walked out with her two plants, two dollars lighter.

Eleven

"The journey of a thousand miles begins with one step."

. . . .

- Lao Tzu

The World Is a Book. Travel Turns the Pages

One of the "perks" of a university professor is the opportunity to teach or conduct research away from the university for an extended period. Most universities have a formal sabbatical program, the purpose being to allow professors to gain knowledge from another university, in the States or abroad. For those outside the university, it sounds a bit like a vacation. However, the idea is to have the professor return with different, and at times better, approaches to research and to share new ideas with students and faculty. While students today are smarter than I ever was, they are still unworldly. They benefit immensely from a teacher who can share new ideas or knowledge of other parts of the world. Interestingly, only a small percentage of professors take advantage of an overseas experience; many do not use it at all. Others stay home to work on a special project like a book. That is sad, and very shortsighted. Going to another part of the country or the world is problematic for the family, to be sure. Schools have to be arranged, housing found, and funds raised to travel, but in the long run, it is worth all the effort. Professors and students are, without doubt, richer for it.

I was fortunate to be invited to New Zealand by Dr. Ian Warrington for a six-month sabbatical. We packed up the entire family and arrived in Palmerston North in April 1988, the beginning of their winter. Jonathan walked to his grade two class with his Maori friend, Taka. Heather was in middle school, and Laura in high school. It was a marvelous experience for the family. It is true that Heather bemoaned the fact we ruined her life by taking her away from her friends, and yes, Laura quickly learned that Kiwis liked their beer. Our home was small, and we found that we didn't need much to live on at all. An old car and a couple of bikes were sufficient. In New Zealand, we learned about recycling; they were decades ahead of us, and we soon became little Kiwis by having more recyclables than garbage. It was also the first time that I saw debit cards used at the gas pump, a development that would take another ten years to be common at home. We traveled throughout the country during holidays and weekends—from Auckland in the North to Milford Sound in the South, and dozens of places in between.

New Zealand should be on everyone's "bucket list." The people are wonderful, the education system top notch, and a picture postcard awaits around every corner. However, I would recommend going during their summer. It is bloody cold in the winter.

I was able to work in state-of-the-art facilities and conducted research on two new crops, *Tweedia* and *Tibouchina*. I also did some teaching at Massey University in Palmerston North and found that students everywhere are the same. As a horticulturist, I learned so much, as did my family. We shared our experiences with friends, colleagues, and students. As a family, we returned home with a renewed commitment of making do with less and succeeded—for about two weeks—then drifted back to the status quo. I managed to bring back a few plants, but most could not be established here.

A large full *Tibouchina* hedge surrounded our home in Palmerston North, and I successfully brought back a cutting, rooted it, and planted it in the trial garden. It comes back every year, sports large purple flowers, and never have I seen a disease spore or insect bothering it. I named it *Tibouchina* 'Athens Blue'. While the plant name never gained a lot of traction, hundreds of cuttings have been rooted over the years, and I often see it under any number of names. So much to see, so little time. The world is like a book, but we must travel if we wish to turn the pages.

In April 1996, I jumped at the opportunity to work with Dr. Julie Plummer at the University of Western Australia in Perth. By this time, the girls were in college, so Susan, Jon, and I winged our way to the farthest western reaches of the continent. Most people visiting Australia never get to Perth. It is over 5,000 miles by car from Sydney and over 11,000 miles from home. Out of the way, but an absolute jewel. In his Perth high school, Jon became much involved in the Australian sporting scene. He had developed into a superb baseball player and found that Aussie baseball was available. He participated in basketball, learned Aussie-rules football, and became quite a cricket player. The girls visited for a couple of weeks over Christmas, and each received at least one marriage proposal. We traveled up and down the west coast, but distances were far. We did not see as much of the interior or the East as we would have liked. We quickly understood the term "dry heat," as Perth is an oven during the summer. We also came to understand the term, "Aussie salute," referring to swiping away the hordes of flies that occasionally come up from the desert. However, more than making up for the heat and the flies was the great camaraderie and endless Aussie barbies.

The movement to native plants in Australia was just starting. I was able to work on growth and forcing of a half dozen genera I had never heard of. We rooted, grew, and flowered many unknown (to me) species of *Dampiera*, *Swainsonia*, *Boronia*, and *Chorizma*. We also tried a few I recognized like *Brachycome* and *Helichrysum*. My head was exploding with new plants. We did not believe many would succeed commercially there or in the United States, but the pure act of learning was immensely fulfilling. Six months seems a long time when planning excursions, but it went by in a heartbeat. The research in Perth continued long after I departed, and I brought back many Australian natives, particularly *Scaevola* and *Dampiera*. I was just getting *Scaevola* established in the greenhouse, but I was too late. Proven Winners® had introduced it a year earlier. I believed the striking blue flowers of *Dampiera*, a *Scaevola* relative, would be well worth testing, but it was simply too difficult to establish. I never put it out for consideration. However, we introduced a wonderful new lantana under the name of 'Aussie Rose' that somehow morphed into 'Athens Rose'. It was a big plant, but breeders and landscapers were already caught up in the "short and compact" dogma of plants. It did not go far.

A number of wonderful tangibles resulted from that visit. We made delightful friends, learned about a great country and their people, and I was able to become even more of a plant nerd, learning species absolutely alien to me. The experiences were also priceless for Jonathan, who later returned to Perth to play and coach baseball. We were not sure he would come back.

In 2007, our last sabbatical took us to the Niagara Parks School of Horticulture in Niagara Falls, Ontario. This unique school is a model for preparing young people for careers in horticulture, particularly for positions that require experience, hands-on knowledge, and problem solving. I asked the school's director, Liz Klose, if she would be interested in my joining them for six months. I must have hit her on a good day.

Students of the Niagara Parks School of Horticulture in the Botanical Garden

Once the paperwork and other hurdles were leaped, we found ourselves in a cute little house in Niagara-on-the-Lake, a truly beautiful part of the world. Who knew how many award-winning wines were grown in that area, or that the world-famous George Bernard Shaw festival was held there every year?

The school was fundamentally different from any four-year or two-year school I had experienced. Students worked from eight to five every day and attended classes as scheduled. Their main job was to manage the one-hundred-acre Niagara Parks Botanical Garden, part of the Niagara Parks Commission, annually visited by over one million people. Dozens of beds were planted, the displays trimmed, and the rose garden maintained. The perennial and annual displays had to be properly labeled, the vegetable and herb gardens educational, and the park continuously manicured. The school taught basic courses like plant identification and use, design, and greenhouse management, but the hands-on work of heavy equipment, tree climbing, and irrigation installation were equally important. The program was three full years. Students lived and ate in a residence hall and were paid approximately sixty cents an hour. There were no spring breaks, summer breaks, study breaks, or football weekends. Internships were required, and weekends were workdays. They spent as much time with pruners, shovels, tractors, and planting as they did with books. Most classes were held in low visitation seasons, few in the summer, and the system worked. While it sounds like it would take a small army to balance classes and run a one-hundred-acre garden that hosted over a million visitors, the entire school consisted of only thirty students. Every student was in high demand, and most went on to successful careers in Canada. I knew I would learn as much from them as they from me.

My role at the school was to be a mentor—a post I always wanted but never knew what it meant—and to help teach some classes. The faculty did not need my help; they were some of the most accomplished people I had come across. I became involved as much as I could and was more than happy to share what little experience I could with these young people. Susan and I made many fine friends and drank a good deal of fine wine. We were so impressed, we returned for another stint the following year. As always, I brought back these experiences to my own classroom and the trial gardens.

Each of the sabbaticals further opened my eyes to the diversity of this thing we called horticulture. Each of them made me a better researcher, better teacher, and better plantsman. What pages I was able to turn!

I'll Be the Plant Guy

The interest in new plants continued unabated throughout the 1990s and remains today. Dozens of new genera were being introduced, and even those sweet potatoes were being lauded by the industry. It was obvious the industry was ready to expand its definition of "new" past another new petunia or impatiens. I was always on the lookout for plants that were unusual but could also be easily propagated and look good in a container at retail. This sounds far easier than it is. I was speaking to more groups every year, and I would always add a half dozen "new" plants I had introduced to the trial garden. I believed many were worthy of additional propagation and distribution. I was talking about rather unknown plants like a dwarf cleome, a dwarf Mexican petunia, or an interesting cuphea. Most of the plants were "one-offs," meaning that I did not develop a series of colors in a given taxon. I also did not feel that the plants I was talking about could or should be patented. I had established these plants similar to those of 'Homestead Purple' and 'Margarita'; that is, using my eyes or people bringing me material they felt may have potential. People in the audience were a little frustrated listening to me, because without means of propagation, distribution, and a sales channel, such plants would always be difficult to find.

In 1998, I was approached by a number of growers. They suggested we put together a corporation of national growers and start introducing "Armi" plants nationally. The initial impetus behind this idea came from John Rader and Kerstin Ouellet. When the word went out that a new brand was forthcoming, we received inquiries from a number of growers wanting to be part of it. Starting a corporation involved articles of incorporation, overseas propagation, distribution facilities, virus testing, and growers who believed and would actually put time and money in the program. I was not at all optimistic; so many great ideas start strong, but after a year or two fade away. However, this group was not to be denied, and Athens Select™ was born, with me as "The Plant Guy." That was a good thing, because that is all I was good at. The distribution, propagation, and marketing were all to be handled by EuroAmerican®. The best part of the program was that it was market-driven. It was not some crazed professor telling the consumer to buy these plants because he said they were good, it was the consumer asking for tough plants and now having something to try. And it worked.

Athens Select Brochure, 1999

The first licensees included Davis Floral, Georgia; EuroAmerican® Propagators, California; Fahr Greenhouses, Missouri; Sunbelt Greenhouses, Georgia; and Wenke Corporation, Michigan. Over the next ten years, licensees changed, but what extraordinary people I worked with:

- Pat Bellrose, Fahr Greenhouse
- Robert Dupont, Dupont Nursery
- Lad Buckley, Buckley Greenhouse
- Earl Robinson, Meadowview Growers
- Sheri Markowitz and Dave Dagen, Emerald Coast Growers
- Aaron Macdonald, Botany Lane
- Ken and Leah James, James Greenhouse
- Robin, Ann, and Robert Davis, Davis, Davis Floral

Many others provided their input and time for an endeavor we all believed in. We brought dozens of extraordinary plants to the industry. When I look through the incomplete list, I marvel at the breadth and scope of the selections we introduced:

- *Acalypha* 'Bourbon Street'
- *Althernanthera* 'Gail's Choice'
- *Cleome* 'Linde Armstrong'
- *Coleus* 'Velvet Lime', *C.* 'Red Ruffles', *C.* 'Mariposa'
- *Cuphea* 'Plum Mist', 'Firefly'
- *Gaillardia* 'Georgia Sunset', *G.* 'Georgia Yellow'
- *Graptophyllum* 'Tricolor', *G.* 'Chocolate'
- *Hypericum* 'Tricolor'
- *Iochroma* 'Purple Queen'
- *Lantana* 'Athens Rose', *L.* 'Lavender Popcorn', *L.* 'New Gold'
- *Pennisetum* 'Prince', *P.* 'Princess', *P.* 'Princess Caroline', *P.* 'Princess Molly'
- *Pentas* 'Stars & Stripes'
- *Plectranthus* 'Variegatus', *P.* 'Athens Gem'
- *Ranunculus* 'Susan's Song'
- *Rosemarinus* 'Athens Blue Spires'
- *Ruellia* 'Ragin' Cajun'
- *Scabiosa* 'Lemon Sorbet'
- *Strobilanthes* 'Persian Shield'
- *Verbena* 'Ron Deal', *V.* 'Homestead Purple'

Do not get the impression that I was responsible for all of these. All the people involved in the program suggested and evaluated, then grew and marketed them. I simply gave them my seal of approval. The program was slowly building up steam. However, as time went on, it was obvious we would never match the marketing power of the big companies, nor were we large enough to compete for space at the box stores. We did not patent many of our earlier introductions. Unfortunately, they soon found their way into other programs. As a result, we received none of the royalties. Times were definitely changing. We recognized that sharks were in the water, just waiting to pick off the unprotected. It was frustrating and sickening to all of us, and when our successful grass program was devoured by others in the late 2000s, it was the last straw for me. It was a bold experiment between an academic institution and the floriculture industry and one of the most unique endeavors of its day. My impending retirement seemed like a good time to phase the program out, and after a thirteen-year run, we closed our doors in 2011. We had a ball.

Would I do it all over again? Absolutely! The camaraderie was the best part—such friendships and respect shone through both good times and bad, and we had plenty of both. Would the other brave members of Athens Select do it again? I don't know, but next time we see each other, we shall raise a glass to Athens Select, the little engine that could.

Nobody Even Grows Them

As a young whippersnapper of a professor, I was ready to do anything. I wanted to do everything. Such is the confidence of the young. I was teaching, working on at least three research projects, selecting plants every now and then, and trying to find funding for the Trial Gardens. In 1984, I made a summer trip to Holland and, while there, I had an epiphany. I saw small acreages everywhere dotted with outdoor cut flowers. I walked through and learned that many were rotation crops for bulb farmers. I saw snapdragons, delphiniums, statice, dianthus, and a dozen more annuals and perennials being farmed, harvested, and sold at the Dutch auctions or farmers markets. Once I saw bunches of eustoma or corn cockle in buckets in the cooler, I was hooked. I went home and asked why we could not do the same in this country?

In no uncertain terms, I was told why not. There was no market. "Don't you know that all cut flowers sold in this country came from overseas?" I was constantly asked, "Why bother with research on cut flowers when nobody even grows them?" In short, they said nobody would care. For sure, the big four—roses, carnations, mums, and baby's breath—were grown overseas in greenhouses, but what about all the other plants I saw moving through the auction? Why could we not explore the possibilities of growing and selling annuals like celosia, snapdragons, and sunflowers, or perennials such as coneflowers or globe thistles? Would not information on bulb crops like tuberose or woody plants such as hydrangeas gain a following? Answers from my colleagues were no, no, no, and no.

Regardless, the next year I put out some research plots of celosia, sunflowers, and coneflowers. Data were collected, and at speaking engagements in 1985, I shared some of the research we were conducting. In the February 1986 issue of *Greenhouse Grower* magazine, I wrote a column showing our plots and outlining my vision. People started to call and wanted to visit. Frankly, I was as surprised as I was pleased.

'Coronation Gold' yarrow in cut flower plots, 1986

I was also tired. To add insult to injury to the comments of the naysayers, I quickly realized that a huge amount of time and effort was needed for maintenance and data gathering. In 1985, I hired a bright young Iowan named Judy Laushman. She had accompanied her husband to Athens in order for him to complete his doctoral studies in botany. Thank goodness she was even more energetic than I. She spent many hot days and long weeks planting, maintaining, cutting, measuring, and bunching field-grown cut flowers. The plots expanded, and so did the data. One day, we decided to answer the question that had been close to the surface since I had started this work: "Did anyone really care?"

We organized the first Field Grown Cut Flower Conference in Athens, Georgia on May 28, 1987. We hoped to attract fifty or so people from the Southeast. If seventy-five had shown up, we would have been thrilled. To our shock, 163 people from California, New York, Maryland, Texas, Ontario, Hawaii, and Washington showed up. Our first question was, "Who are all these people?"

Today 163 people may not sound like much. But for a topic not even on the radar, it was obvious that people did care. In August 1988, thirty people interested in the business of cut flower growing called a meeting in Chicago to look into forming a new association. And so they did. The Association of Specialty Cut Flower Growers (ASCFG) was born, and in the ensuing years has built a loyalty base like no other.

The association produces one of the finest periodicals for commercial growers and hosts national and regional meetings. It conducts the only cut flower seed trial in the country. The ASCFG also sponsors academic and grower research, offers a scholarship program, and manages nearly one thousand members. I think of the associations that have come and gone with far more money and a far larger audience, yet the ASCFG is still going strong twenty-seven years later! I am immensely proud, but its success has had nothing to do with me. Judy Laushman's leadership and creativity has taken this little organization to one of the finest in the country. It is not the largest or the loudest, but if you meet any of the long-term members of ASCFG, you will agree they are the most passionate.

Making Money on Company Time

Publications have long been the backbone of any successful academic. In the sciences, refereed journal articles are the world currency; other publications such as *Greenhouse Grower, GrowerTalks*, or *Fine Gardening* magazine are nice to have but mean little to those evaluating your performance. For professors in languages, the arts, and history, books are very important, but not so in the sciences—at least they were not at UGA. While a textbook or serious reference tome was given consideration, a book that would be sold in reasonable numbers to lay people was low on the ladder. This negative perception was rooted in the fact that books are not refereed, thus not particularly valid in the world of science. This is one reason you will see few horticulture/gardening books written by university professors.

A book that was highly successful and could not be ignored, even though it was as popular with lay people as with academics, was Dr. Michael Dirr's *Manual of Woody Landscape Plants*. When I arrived at UGA, he was already working on the third edition. It was used as a textbook, as well as a source of enjoyment and learning among gardeners and professionals. It was one of the first books to span the two camps.

In 1985, my first book, if you can call a forty-four-page grower manual a book, was published about growing seed geraniums. It was the first in a series of grower manuals with Timber Press in Portland, Oregon. I continued doing the series with authors from other universities, but I wanted to "be like Mike." I wanted to write a book that was a reference manual, an enjoyable read, and of value to gardeners and professionals alike. I also hoped it would also be useful as a college text. In 1989, the first edition of *Herbaceous Perennial Plants* appeared, 646 pages long. Michael not only encouraged me, but he also published the book under his new company, Varsity Press. The book took over a year to write, but I was heartened by the positive reviews and healthy sales.

People took notice that there were two successful authors in the department pandering to the lay audience. It was not so much that we were writing books—many of our colleagues were quite successful with books of their own—but somehow people believed we were making lots of money. Believe me, it was a perception only. However, a few others viewed it as a conflict of interest. We were being paid by the taxpayer, and our time should be spent doing only those things we were hired to do: teach and perform research. We shouldn't be working on books, at least not on books that would make some money. That a scientific paper might be read by one hundred colleagues, while a good book might be read by ten times that many, did not seem to hold water in their argument. We were "making money on company time," and while we were not asked to stop, our department chairman asked that we do it on our own time.

If "they" had asked me about working during school hours, I could have told them that I wrote every morning at five. I was too busy at school to find even an hour to work on a book. I wanted to be with the family when I came home from work, not steal time from them in the evening. I still adhered to Dr. Ormrod's advice about working an hour a day; my hour simply occurred early in the morning. The issue went away over time. We accomplished all the university asked of us and enhanced the reputation of the department of horticulture among professionals and lay people with a few good books to boot.

Herbaceous Perennial Plants is now in its third edition and weighs in at 1,109 pages. I have also managed to write thirteen other books in my time, some quite good, some forgotten, but all penned at five in the morning.

Twelve

*"An autobiography is an obituary in serial form
with the last installment missing."*

. . . .

- Quentin Crisp

I've Never Even Heard of the Hustle

When you get a little notoriety, people ask for your time. I have been asked to speak all over the country, give classes here and there, do radio shows and webinars. While I am pleased to share what I know, deep down I am not as altruistic as I appear. I may charge an honorarium and even if I do not, I realize I may get additional book sales in return. However, I have done my best to give back to this industry when I can. I served on boards, gave freely of my advice when asked, and volunteered on many occasions. I just never thought I would be asked to dance.

Project Safe raises awareness of violence against women in many parts of the country. They do amazing work through support services, safe houses, and legal help, all of which require funds. The Athens chapter was extraordinarily creative. Starting in 2008, they organized an annual "Dancing with the Stars" program loosely based on the hit television show. They raised thousands of dollars by promoting the program using Athens-area celebrities and talented dance instructions or dance students.

Family Fans for Allan's Dancing Debut, 2010

In 2010, I was asked to be one of the celebrity dancers. My students, friends, and family thought that Armitage dancing on the big stage was a real hoot! I had made a fool of myself many times before, but this was the first time I volunteered to do it in front of an audience.

The goal was to let your friends know about the upcoming competition and have them vote for you. Each vote cost one dollar. Once I accepted, I let my horticultural buddies know I wanted their support—my students were even expected to put in a dollar from their beer money—and come to the show. During the show, the audience could also vote. I may fall on my keister, but I would raise some money for Project Safe.

. . . .

I met my dancing partner Liza, a dance student at UGA. She was amazing! Liza is athletic, graceful, and wonderfully patient with an oaf with two left feet. She looked me over and said, "We are going to do the Hustle." I knew the waltz, the jitterbug, had heard of rumba and samba, but the hustle? "What is that? I've never even heard of the Hustle."

Things only got worse when she said, "We're going to dance it to 'Shake Your Groove Thing,' by Peaches and Herb. Liza looked at my blank expression. "Not to worry, I'll show you." For the next eight weeks, she did just that. Every step was choreographed; every whirl, twirl, and jump was timed to the music. It was the hardest thing I ever did. I had never felt quite as uncoordinated as I did in learning these steps. After I finally stopped crushing her feet or tripping over my own, she looked at me and said, "We are going to win this thing. By the way, we have to talk about your costume."

. . . .

John Travolta may have been better looking, a sharper dresser, and a better dancer than me, but he didn't have my hat. Eleven other celebrities and their partners appeared at the Classic Center that evening. A crowd of approximately two thousand people settled in, buzzing in anticipation of the fun. We were the ninth couple of the evening, and everyone was in great spirits, cheering on their favorites and making sure that Project Safe was doing well. My students, friends, and grandkids sported a green Couple Number Nine T-shirt. They yelled, screamed, and carried on. At the beginning of the routine, I took off the hat and joyfully sent it into the audience, like a boy throwing a Frisbee. People went crazy.

At the end, the judges gave their opinions to the spectators and, while all were good, one of them said, "I would have graded you higher, but your hat injured a lady in the third row." More laughs, more good times. We came in fourth (poor judging), but finished third for most money raised. The dancing was great—it is still a big hit on YouTube —but the best thing is, between all of us, we raised a record $96,000 for Project Safe. Next time you see me, I'll show you the Hustle .... in your dreams.

Allan Travolta
Dancing with the Stars

It's Not Just About the Hat

What can I say about this crazy hat? Like the Aflac® duck and the GEICO® gecko, it is its own icon. Walking through a crowded trade show without the hat is like a disguise in reverse; without it, I am invisible. I have been told many times, "I didn't recognize you without the hat." One woman said without hesitation, "You are more handsome with your hat on." Another complains that my handsome face is hidden by the frying-pan-sized brim. If I want to be sure I am recognized, I only need don the *chapeau*. I have no idea how the hat took on a life of its own, but I know why I adopted it.

Georgia can be a hot place. The summer sun is brutal and unforgiving and is no friend to a person with light skin. As a boy, I suffered bad sunburn one summer, severe enough to result in third-degree burns. In short, I was a lobster. In those days, the bronze look was highly sought after. Few people understood the detrimental long-term effects of sun exposure. You would think that once I arrived in Georgia—I would have known better, but even then—I seldom wore a hat. Baseball hats were for playing baseball. I never thought of wearing one, even though it was obvious my students at Georgia slept in them. I wore a golf visor for a year or two, but that did little but shade my eyes.

After introducing me to the joys of liquid nitrogen to remove pre-cancerous cells on my face, my dermatologist told me that my life would be prolonged if I kept the sun off my face. "Cover your head!" I tried all sorts of hats, but I simply was not a hat person. I would often leave the house without one.

One day, I saw a canvas hat at an outdoor store from a company called Tilley®. The hat was obviously developed by someone with a sense of humor, having "Instructions for Use" and humorous comments from fans all over the world. It was expensive, but it was comfortable. I bought my first Tilley hat in the mid-1990s. When I discovered the wide-brimmed model of the hat, I latched on to it because it really did shade my face. It is not at all a stylish hat. Those who know me well have rolled their eyes on many occasions. Like a reformed smoker, I became somewhat of an evangelist on wearing a hat outside. At the beginning of every course, I asked that no hats be worn in the classroom, but they were mandatory outside. When we gathered in the trial gardens or elsewhere, the first time a student did not bring a hat, I was an understanding fellow. I simply stated that if he or she forgot again, he or she would have to write a ten-page essay on skin cancer. In all the years I taught, I only had one essay cross my desk.

It is only a hat, for goodness sake, but I must admit it has been a blessing. I am not sure what will get me in the end, but if it is sun-induced melanoma, it won't be without a fight. I have been spot-treated with so much liquid nitrogen that I should have bought shares in the company. I believe that wide-brimmed thing has added years to my life, and I have recommended it to many others. The hat has been with me so long that people even associate my "notoriety" with it, to which I reply, "It's not just about the hat."

Retirement Does Not Mean Inertia

I retired from the University of Georgia in 2013. I wanted to go out on top of my game. The thought of being one of those professors shuffling around the halls because they had nothing else to do terrified me. Not that such a thing would have occurred; I was still enjoying the students and they me, and some new crop introductions like 'Misty Lace' goats beard and 'Iron Butterfly' ironweed were doing just fine. The trial garden was running on all cylinders. We were publicizing good plants and ignoring bad ones.

We were also involved in landscape roses and ornamental vegetables in a big way. I had excellent relations with growers, landscapers, and retailers and was working with most of the world's breeders; in short, all was well. I was also fortunate in having many hobbies; I can spend hours in my little garden or easily get lost in a good book. I play the guitar, poorly enough so that Susan closes the door, but

I play with gusto. I enjoy squash and tennis and have recently become a bicycle rider, covering many miles a week on my delightful green bike.

If I have missed anything from the university, it is the interaction with the students. I have lost count of the many fine young people who sat in my classes, worked in the gardens, or assisted me in research. Every one of them in some way made me a better person. We learned and we laughed together, and there is seldom a place I visit that one of my students does not come up and say, "Hi Dr. A, remember me?"

Inertia has not yet set in. I am still very much involved in horticulture. You will still see me presenting lectures at garden and trade shows, telling stories, or talking plants. I will continue to do so until I am no longer in demand. My little travel company, Garden Vistas, is still satisfying my travel bug. For the last twenty-five years, my incredibly talented colleague, Linda Copeland, and I have organized trips for gardeners and professionals to the Great Gardens of the World. I teach. I laugh. I walk and drink wine from Stockholm to Sydney. *Condé Nast* it is not, but we have a ball while learning about plants, people, and the world in general. Maybe you will join me some day.

I am also spending time teaching online classes and have even developed an app for smart phones and tablets. All these activities help to keep me off the street.

I am managing to spend more time with companies and people I have always admired, and equally fortunate that I can say no to those I don't. The people I work with are like Fortune 500 companies, admired and highly sought after. *Greenhouse Grower* has not yet fired me, and what a relationship we still have. The Ball Horticulture® Company—Nature's Source® Plant Food—is one of the finest in the entire field of horticulture. Working with Janet Curry, Amy McCormack, and their team has been a treat. My friend Steve Jarahian continues to take the technology of soilless mixes to new heights at Oldcastle® Company, makers of Jolly Gardener® soils. I enjoy staying involved with the best plants, so I have become associated with Randy Hunter and Maria

Zampini at the HGTV HOME® Plant Collection. And what good times I have had working on the Treadwell and Deer-Leerious programs with all the people at the Perennial Farm in Glen Arm, Maryland.

I do so not only because I believe in their products, but because I so admire the people involved. Anyone can find a good product. Finding great people is far more challenging. I will stick with them as long as they will have me.

And on that note, I must add a word or two about Rick Watson and Ed Kiley, the two gentlemen I have been working with at the Perennial Farm. It is because of them that this book has been penned. We have enjoyed many stories and many good times over a few beers and Maryland crabs smothered in Old Bay seasoning. So many stories in fact, that they talked me into writing about myself. Without them, I guarantee this book would not have been written. If you have enjoyed this journey, thank them. If you completely wasted your time and money, I will send you their emails.

The Perennial Farm family
Steve Bothwell, Ed Kiley, Mark Huber, Rich Poulin, Gail Watson,
Tom Watson, Rick Watson, Alice Tomasello, Diane Holhubner

As Susan and I got older, so did the kids. Susan pursued her degree in nursing, and for many years managed the Occupational Health Clinic for USDA in town. We walk, ride, and travel together. She is still patient.

Laura topped out at 6'1" and is in charge of a speech pathology unit in a large hospital in Macon. She and her husband Ray have two beautiful children, both of whom will also be as tall, graceful, and smart as their mother.

Heather was the wild girl as a teen. Today she and her husband David have four children, including a set of twins, all within five years of each other. Susan and I smile: "Yes, there is a God." She balances these wonderful kids with her job as a diabetic educator at the Medical College of Georgia in Augusta.

As for Jonathan, he grew as well, topping out at 6'5". He followed his dream and was drafted by the San Francisco Giants and played in their system through Triple A, getting a taste of "The Bigs" in spring training. As such things happen, he did not stay in the majors, but went

Back row: Heather, Laura, Jonathan, Susan, Allan
Front row: Drew, Ben, Hampton, Kate, Will, Mary Grace

back to school and obtained his master's degree in economics. Because of his athletic experiences and mathematical skills, he was recruited by a start-up company in San Francisco and loves it. He and his wife Mandy are living the good life on Russian Hill.

The best thing about all this is that, regardless of the journeys the children have taken—jobs, kids, dogs, and moving—they are still very close to each other, and to their parents. It simply doesn't get any better than that.

If you see me walking at a trade show or down the street, please stop and say hello. I'll be the one in the hat wearing chambray.

In Conclusion

Upon retirement, I was honored by friends and colleagues for my efforts in the Trial Gardens at the University of Georgia. The gardens had come a long way in thirty years, and people spoke fondly of their many visits there. They left no doubt that what was created had positively influenced hundreds of people. I was proud, humbled, and surprised!

Unbeknownst to me, one of the speakers who came forward that morning was my wonderful and beautiful daughter Laura. She and her children, Hampton and Mary Grace, had driven a considerable distance to be with me that day. I was touched that she came, but especially touched by her comments.

"It's not the recognition that keeps this man going. It's the sharing of his love of the dirt, flowers, and seeds, the stories he can tell to make a connection with these. It's the smiles, the jokes, and the laughter that let my dad know there's more to his legacy than just happily ever after."

– Laura Yarbrough

Pictured left:
Allan, Susan, Laura, and her children,
Hampton and Mary Grace

Near the end of her remarks, with everyone absolutely rapt, she stated that one of the credos in her life was something I told her many years ago: "Dig a ten-dollar hole for a ten-cent plant." I remembered saying those words, but had no idea they had stuck with her. She then mentioned how those words helped her out by saying, "You never know how things will work out, and being ready for anything keeps small problems small."

People smiled. I wiped a few tears from my eyes. After all, it was allergy season.

More Titles by
Dr. Allan Armitage

Publications can be ordered from Dr. Allan Armitage's website. Dr. Armitage will personally sign each book. They make a great gift for the horticulture student or avid gardener in your life! Visit: http://www.allanarmitage.net/shop/

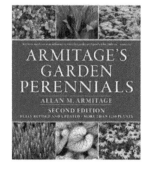
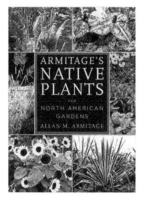

**Herbaceous
Perennial Plants**
A Treatise on Their
Identification, Culture
and Garden Attributes
Third Edition

**Armitage's Garden
Perennials**
*Second Edition,
Fully Revised
and Updated*

**Armitage's
Native Plants**
for North American
Gardens

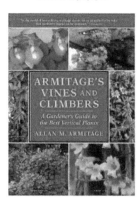
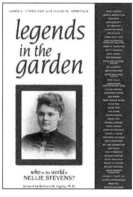
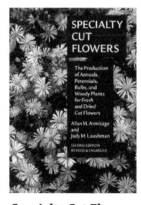

**Armitage's Vines
and Climbers:**
A Gardener's Guide
to the Best
Vertical Plants

**Legends in
the Garden**
Who in the World
is Nellie Stevens?

Specialty Cut Flowers
The Production of
Annuals, Perennials,
Bulbs, and Woody
Plants for Fresh and
Dried Cut Flowers

Take Online Classes with Dr. Allan Armitage

Armitage's Herbaceous Sun and Shade Perennials Online Courses

- Self-study certificate courses for green industry and gardening audiences
- Professional development/staff training programs authored by Dr. Armitage
- Offers best landscape and gardening use recommendations
- Between eighteen and twenty genera studied in each course

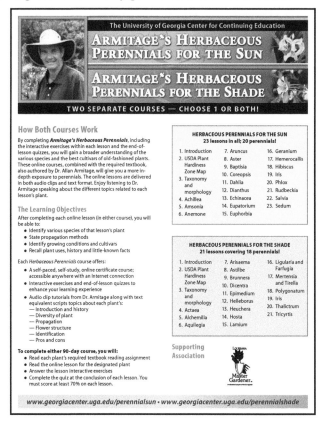

Enroll Anytime. Register Now!

www.georgiacenter.uga.edu/perennialsun
www.georgiacenter.uga.edu/perennialshade

The University of Georgia
Center for Continuing Education
1785

Email: questions@georgiacenter.uga.edu
Phone: 800.811.6640 or +1.706.583.0424

Dr. Allan Armitage Goes Digital

Dr. Allan Armitage has developed a garden app for Android and iOS platforms. **Armitage's Greatest Perennials & Annuals** app spans the gap between the horticultural industry and consumers. Have the plant expert in your hip pocket.

Dr. Armitage selects his favorite annuals, perennials, ornamental grasses, and vegetables. The app is continually updated with plants and garden centers on a monthly basis.

Dr. Armitage lets readers know why he chose particular plants and cultivars. Allan provides hints for success in the home landscape. Users can filter for annuals or perennials and select sun, shade, or USDA zones.

Deer Resistant Plant Suggestions

A unique and popular feature is the deer browsing rating for all plants.

Numerous photos and videos supplement the plant-specific information.

Plants for Sun and Shade

Dr. Allan Armitage lists his favorite plants to use in sun and shade situations.

The app is great to have on hand when you're in a garden center. It's your personal plant shopper!

Retail Garden Centers

Find a plant you can't live without?
Dr. Allan Armitage lists independent garden centers by state.

If you are a plant geek, you can locate garden centers that carry the unusual.

Buy the App Today!

Only $4.99 on Google Play and Apple iTunes stores.

Learn more at: www.allanarmitage.net